Alternative Energy

DeMYSTiFieD®

DeMYSTiFieD® Series

Alternative Energy

DeMYSTiFieD®

Stan Gibilisco

Second Edition

New York   Chicago   San Francisco   Lisbon   London   Madrid   Mexico City
Milan   New Delhi   San Juan   Seoul   Singapore   Sydney   Toronto

The **McGraw·Hill** Companies

Cataloging-in-Publication Data is on file with the Library of Congress.

McGraw-Hill books are available at special quantity discounts to use as premiums and sales promotions, or for use in corporate training programs. To contact a representative, please e-mail us at bulksales@mcgraw-hill.com.

Alternative Energy DeMYSTiFieD®, Second Edition

1 2 3 4 5 6 7 8 9 0   DOC/DOC   1 2 0 9 8 7 6 5 4 3 2

ISBN 978-0-07-179433-6
MHID 0-07-179433-6

Sponsoring Editor
Judy Bass

Acquisitions Coordinator
Bridget L. Thoreson

Editing Supervisor
David E. Fogarty

Production Supervisor
Pamela A. Pelton

Project Managers
Nancy Dimitry,
Joanna Pomeranz,
D&P Editorial Services

Composition
D&P Editorial Services

Copy Editor
Nancy Dimitry,
D&P Editorial Services

Proofreader
Don Dimitry,
D&P Editorial Services

Art Director, Cover
Jeff Weeks

Cover Illustration
Lance Lekander

To Samuel, Tony, and Tim

About the Author

Stan Gibilisco, an electronics engineer, researcher, and mathematician, has authored multiple titles for the McGraw-Hill *Demystified* and *Know-It-All* series, along with numerous other technical books and dozens of magazine articles. His work has been published in several languages. He maintains a Web site at **www.sciencewriter.net**.

Contents

Introduction

This book can help you learn or review the fundamentals of conventional and alternative energy technology without taking a formal course. It can also serve as a supplemental text in a classroom, tutored, or home-schooling environment. You might even get some ideas for upgrading your home or business.

How to Use This Book

As you take this course, you'll find an "open-book" multiple-choice quiz at the end of every chapter. You may (and should) refer to the chapter text when taking these quizzes. Write down your answers, and then give your list of answers to a friend. Have your friend tell you your score, but not which questions you missed. The correct answer choices are listed in the back of the book. Stay with a chapter until you get most of the quiz answers correct.

The course concludes with a final exam. Take it after you've finished all the chapters and taken the end-of-chapter quizzes. You'll find the correct answer choices listed in the back of the book. With the final exam, as with the quizzes, have a friend reveal your score without letting you know which questions you missed. That way, you won't subconsciously memorize the answers. You might want to take the final exam two or three times. When you get a score that makes you happy, you can (and should) check to see where your strengths and weaknesses lie.

I've posted explanations for the chapter-quiz answers (but not for the final-exam answers) on the Internet. If you enter "Stan Gibilisco" as a phrase in your favorite search engine, you should get my Web site as one of the first hits. You'll find a link to the explanations there. As of this writing, the site location is

www.sciencewriter.net

Strive to complete one chapter every two or three weeks. Don't rush, but don't go too slowly either. Proceed at a steady pace and keep it up. That way, you'll complete the course in a few months. (As much as we all wish otherwise, nothing can substitute for "good study habits.") After you finish this book, you can use it as a permanent reference.

I welcome your ideas and suggestions for future editions.

Stan Gibilisco

Alternative Energy
DeMYSTiFieD®

chapter 1

Heating: The Basics and the Primitives

Burning "dead plant matter" directly can sometimes make the difference between comfort and freezing. Some people can obtain wood, corn, and coal more easily than they can get access to conventional fuels (such methane, propane, or oil), or alternative energy sources (such as wind, solar, or geothermal).

CHAPTER OBJECTIVES

In this chapter, you will

- Compare various units for measuring heat energy.
- Contrast energy with power.
- Discover the three different modes of heat transfer.
- Learn how to burn wood logs, wood pellets, corn kernels, and coal to heat your house.
- Enumerate the assets and limitations of wood, corn, and coal as energy sources.

Energy, Power, and Heat

Have you heard the terms *energy*, *power*, and *heat* used interchangeably, as if they mean the same thing? They don't! Energy is power manifested over time. Power is the rate at which energy is expended. Heat is any form of energy transfer that causes changes in temperature. We can express energy, power, and heat in several ways, and they all can occur in various forms.

Joules, Watt-Hours, and Kilowatt-Hours

Physicists measure and express energy in units called *joules*. One joule (1 J) represents the equivalent one *watt* (1 W) of power expended, radiated, or dissipated for one *second* (1 s) of time. A joule works out as the equivalent of a *watt-second*, and a watt works out as the equivalent of a *joule per second*. Mathematically:

$$1\ J = 1\ W \cdot s$$

and

$$1\ W = 1\ J/s$$

In electrical heating systems, you'll sometimes encounter energy units known as the *watt-hour* (symbolized W · h or Wh) or the *kilowatt-hour* (symbolized kW · h or kWh). A watt-hour represents the equivalent of 1 W dissipated for 1 h, and 1 kWh represents the equivalent of one *kilowatt* (1 kW) of power dissipated for 1 h, where 1 kW = 1000 W. Therefore

$$1\ Wh = 3600\ J$$

and

$$1\ kWh = 3{,}600{,}000\ J$$
$$= 3.6 \times 10^6\ J$$

Calories and Kilocalories

Once in awhile, you'll encounter the *calorie* as a unit of heat measure. One calorie (1 cal) equals the amount of energy transfer that raises the temperature of exactly one gram (1 g) of pure liquid water by exactly one degree Celsius (1°C), if none of the water vaporizes in the process. A calorie also represents the amount of energy lost by 1 g of pure liquid water if its temperature falls by 1°C, if none of the water freezes in the process.

The foregoing definitions of the calorie hold true only if the water remains liquid during the entire process. If any of the water freezes, thaws, boils, or condenses, the definition no longer works. At standard atmospheric pressure on the earth's surface, this definition holds true for temperatures between approximately 0°C (the freezing point of water) and 100°C (the boiling point).

TIP *The* kilocalorie *(kcal), also called a* diet calorie, *equals the amount of energy transfer involved when the temperature of exactly 1 kg, or 1000 g, of pure liquid water, rises or falls by exactly 1°C without any of it changing state (vaporizing or freezing) in the process. As things work out, 1 cal = 4.184 J, and 1 kcal = 4184 J. If someone tells you that a slice of bread "contains 75 calories," she means that if you burn it up until all the heat energy has been released, you'll end up with 75 kcal (not 75 cal) of additional heat energy in the surrounding environment.*

British Thermal Units (Btu)

In the United States, home heating product vendors often quote an energy unit called the *British thermal unit* (Btu). You'll hear or read about Btu (or "Btus") in ads for American furnaces and air conditioners.

One British thermal unit (1 Btu) equals the amount of energy transfer that raises the temperature of exactly one pound (1 lb) of pure liquid water by exactly one degree Fahrenheit (1°F). It's also the amount of energy lost by 1 lb of pure liquid water if its temperature falls by 1°F. This definition, like that of the calorie, holds true only if the water remains in the liquid state during the entire process.

If someone talks about "Btus" literally, in regards to the heating or cooling capacity of a furnace or air conditioner, she's using the term improperly. She really means to quote the rate of energy transfer in *British thermal units per hour* (Btu/h), not the total amount of energy transfer in British thermal units. We express the real-world heating ability of a stove or furnace in terms of power, not energy. As things work out,

$$1 \text{ Btu} = 1055 \text{ J}$$

in terms of energy. For power, we have

$$1 \text{ Btu/h} = 0.293 \text{ W}$$

and

$$1000 \text{ Btu/h} = 293 \text{ W}$$

Conversely,

$$1 \text{ W} = 3.41 \text{ Btu/h}$$

and

$$1 \text{ kW} = 3410 \text{ Btu/h}$$

TIP *A home furnace with a heating capacity of 100,000 Btu/h operates at the equivalent of 29.3 kW. That's roughly the amount of power consumed by 20 portable electric space heaters operating at full blast!*

Forms of Heat

If you place a kettle of water on a hot stove, heat migrates from the burner to the water. This phenomenon constitutes an example of *conductive heat*, also called *conduction*, as shown in Fig. 1-1A. When an *infrared* (IR) lamp, also called a "heat lamp," shines on your sore shoulder, energy migrates to your skin's surface from the filament of the lamp, an example of *radiative heat*, also called *radiation*, as shown in Fig. 1-1B. (The heat then gets into the ailing joint by conduction in body tissues.) When a fan-type electric heater warms a room, air passes through the heating elements and circulates into the room where the hot air rises and mixes with the rest of the air, raising the overall air temperature. That's an example of *convective heat*, also called *convection*, as shown in Fig. 1-1C.

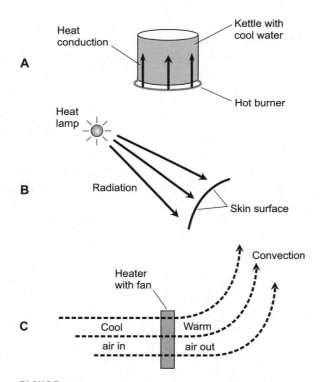

FIGURE 1-1 • Examples of heat energy transfer by conduction (A), radiation (B), and convection (C).

The Wood Stove

Wood fuel has served humankind since the stone age! *Wood stoves* have become sophisticated in recent years, with the advent of optimized air intake systems, blowers, thermostats, and *catalytic converters* similar to the emission-control devices in motor vehicles.

How It Works

In a wood stove, a controlled fire heats a heavy cast-iron box, which in turn emits heat in the form of IR radiation. This radiant energy warms the walls, floor, ceiling, and furniture. In addition, conduction transfers heat to the air by direct contact with the hot stove and with the warmed walls, floor, ceiling, and furniture. The warmed air rises, causing continuous air circulation (convection) that helps to equalize the temperature throughout the room. A wood stove, therefore, heats a room by all three modes familiar to the physicist (Fig. 1-2).

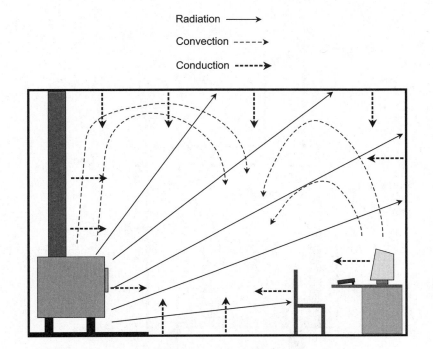

FIGURE 1-2 · A wood stove heats a room by taking advantage of conduction, radiation, and convection.

Figure 1-3 is a cutaway view of a typical wood stove as seen directly from the left-hand side of the box. The *primary air intake* ensures that some air always flows into the *firebox*. The intensity of the fire can be controlled by adjusting the *secondary air intake*. Opening this valve increases the rate of the burn and increases the temperature. Closing it reduces the burn rate to a minimum. The *catalytic converter* changes most of the energy contained in the smoke into usable heat, and also reduces the particulate pollution that goes up the *stack*. In fact, catalytic converters are designed specifically to minimize pollution.

A large wood stove can provide upwards of 150,000 Btu/h of heating power, provided it is kept operating properly. That's a heating capability just about equivalent to the gas furnace that you'll find in a big house.

TIP *A good wood stove can heat a large room in a reasonable time, even when the outside temperature remains far below freezing. If you install the stove in the basement of your house near the main air intake vent for a conventional furnace,*

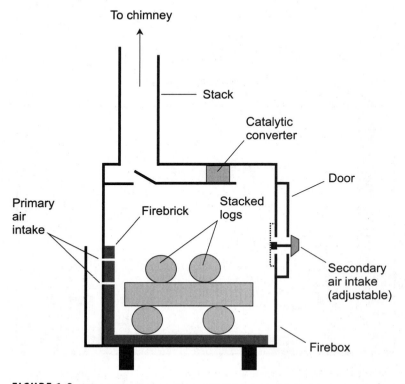

FIGURE 1-3 • Side cutaway view of a contemporary wood stove.

the furnace blower can circulate the heated air throughout the house, even if the furnace isn't producing any heat of its own. In this way, a large wood stove can keep a medium-sized house warm even during Arctic-like cold snaps.

Advantages of Wood Stoves

- A basic wood stove requires no external power source to heat the room in which it's located. If that room is on the lower level, doors can be left open so the warm air will rise to heat the rest of the house and the stove can serve as an emergency heat source when all normal utilities have gone down. You don't need to have a backup battery or generator.

- Wood constitutes a renewable fuel. Trees can be deliberately grown and harvested to provide fuel for heating, just as trees are grown and harvested to provide lumber for building.

- Burning wood in a stove can minimize waste. Wood that would otherwise get burned at a brush dump or create a wildfire hazard (dead wood in a forest, for example) can be gathered, cut, and used to heat homes.

- Frequent and regular use of a wood stove can significantly reduce the cost of heating a home by conventional means such as gas or oil. It can also mitigate the impact of a sudden, severe shortage of conventional fuels—provided, of course, that a source of wood fuel is available at a reasonable price.

- For some people, wood stoves have esthetic appeal.

Limitations of Wood Stoves

- Wood stoves can pose a danger to life and property! Before installing and using one, read the instruction manual. Wear safety glasses, and completely cover yourself (including your hands) with fire-resistant clothing when working around an active wood stove.

- You must acquire and maintain a stockpile of dry, cut wood. Logs must be split and cut to lengths small enough to easily fit in the firebox. All of this preparatory work can prove inconvenient and time-consuming.

- You should allow your firewood to dry for at least 12 months after being cut, and preferably for 18 months. Fresh-cut wood has high moisture content, so it burns inefficiently (and sometimes won't burn at all).

- You can get cut, dried wood from commercial sources in some locations, but it almost always costs a lot of money!

- The fire requires constant attention while the wood stove operates. You should never leave your house with the thing going "full blast."
- Wood has relatively low efficiency as a fuel source. No wood stove can equal the efficiency of a top-of-the-line gas furnace.
- A wood stove requires frequent cleaning. You should allow all the ashes and coals to cool down completely before you remove them, and this precaution translates into stove downtime.
- The chimney needs periodic cleaning to prevent buildup of *creosote*, which can ignite and cause a dangerous *flue fire* (also known as a *chimney fire*).
- Some municipalities restrict the installation and use of wood stoves, and a few places won't let you have one at all. Some insurance companies won't underwrite a policy for a home that has a wood stove, especially if it constitutes the primary (main) or only heat source.
- If you want to heat your whole house with a wood stove, the room containing the stove will become hot; if it's a small room, the temperature can easily rise to over 110°F (43°C).

? Still Struggling

If you want a concise reference book for wood stove operation and wood fuel in general, try to get your hands on a copy of *All That's Practical About Wood* by Ralph W. Ritchie (Springfield, Oregon: Ritchie Unlimited Publications, 1998). But remember: Neither that book nor this one can serve as a safety guide. If you have any doubt about the installation and use of a wood stove after reading its instruction manual, contact your local fire marshal, who will probably want to inspect your system anyhow.

PROBLEM 1-1

What, besides cut firewood, can a wood stove burn to provide heat? How about charcoal, or coal, or flammable liquids?

✔ SOLUTION

Most wood stoves are designed to burn properly cut, dry wood, and nothing else. Charcoal or coal gets too hot. The use of any flammable liquid

can cause an explosion and set clothes, carpeting, and furniture on fire instantly. A few specialized wood stoves can burn coal; but before you try to burn coal in your wood stove, check the instruction manual!

Pellet Stoves and Furnaces

There's a more efficient, cleaner, and safer way to burn wood than the old-fashioned "log pile" method. Sawmills compress waste sawdust into pellets that can burn in *pellet stoves* and *pellet furnaces*.

How They Work

Figure 1-4 illustrates the internal components of a pellet stove. You pour the pellets, which look something like dry pet food, into a *hopper*. A feed system, usually comprising a *motor* and an *auger* or other mechanical device, supplies pellets to the firebox at a rate that you can set manually or automatically, depending on the type of stove and on your preference.

Wood pellets have energy density too great for burning in a free-standing pile. You can't fill up an ordinary wood stove with pellets and expect it to work. In order for combustion to take place, air must flow through the pellet pile. A pellet stove has a *blower* that forces air through the firebox, ensuring combustion. The air can come from outside the house to prevent *negative pressure* that would otherwise draw cold air into the house. The exhaust fumes vent to the

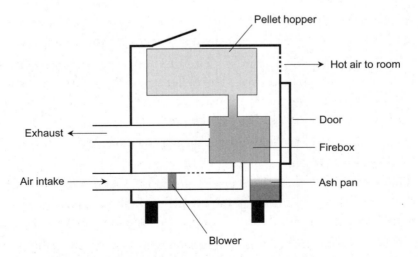

FIGURE 1-4 · Simplified functional diagram of a wood pellet stove.

outside as well. Heated, unpolluted air from inside the stove, after having been warmed by the firebox and a corrugated mass of metal called a *heat exchanger*, flows into the room.

A pellet furnace basically constitutes an oversized pellet stove. A blower forces the hot air into ductwork. Ideally, if the ductwork is properly arranged, the heated air circulates uniformly throughout the house with the help of convection from the lower levels to the upper levels, just as it would do with any other type of forced-air furnace.

TIP   *Free-standing pellet stoves are typically rated at maximum outputs between 30,000 and 70,000 Btu / h. Large pellet furnaces can deliver considerably more heat power, in some cases over 100,000 Btu / h.*

TIP   *You can usually install a pellet furnace directly in place of a forced-air gas furnace with little or no modification to the existing air distribution system in your house. A home heating professional can examine your system and let you know for sure whether or not you can do it.*

Advantages of Pellet Stoves and Furnaces

- Pellet stoves work more efficiently than wood stoves do. The refined pellets contain almost no moisture, little or no pitch (sap), no dirt, no insects, and no bark. As a result, you get more heat, less pollution, and less ash per kilogram of fuel.
- Pellet stoves operate more safely than wood stoves do. The exterior of the pellet stove never gets dangerously hot (except for the door glass).
- With a pellet stove, you can regulate the temperature more easily than you can do with a wood stove. The pellet stove doesn't need constant attention. You can set the thermostat and pretty much forget about the stove, except for periodic hopper refilling.
- You can dispose of the ash without having to endure extended periods of stove downtime.
- Pellet stoves don't need chimneys. The exhaust gases can vent out through the side of the house, in the same manner as high-efficiency gas furnaces work. You don't have to worry about the buildup of creosote in a chimney.
- Pellet stoves or furnaces are often legal in regions or municipalities where wood stoves are forbidden.

Limitations of Pellet Stoves and Furnaces

- If the electrical power fails, a pellet stove won't work unless it has a *backup battery*, or you have a generator that creates a clean, alternating-current (AC) *sine wave*. The blower motor requires electricity to operate, and the stove won't function properly without it.

- Pellet stoves have sophisticated internal electronics. These circuits, which resemble those found in modern gas furnaces, take most of the hassle out of operating the system—until a component fails. Then the whole machine goes down, and you can't use it again until a qualified technician repairs it.

- A pellet-burning stove or furnace should have a *transient suppressor*, also called a *surge suppressor*, to minimize the risk of system failure in the event of a power-line "spike." You should install the transient suppressor, available at most hardware stores for a few dollars, between the utility outlet and the pellet stove.

- If a foreign object gets into the feed system, it will jam, shutting down the stove. If you're away for a day or two and this sort of thing happens, you'll return to a cold house.

- Pellets, while easily available in some locations, are hard to get in other places. You'll have to stockpile them, in much the same way as you stockpile wood for a wood stove.

- Pellets come in heavy bags, usually 18 kilograms (kg) or 40 pounds (lb). In cold weather, you'll have to fill up the pellet hopper at least once a day, and maybe twice. You'll end up lifting and hauling a lot of pellet bags.

PROBLEM 1-2

Can a wood or pellet stove safely vent into the same chimney as another appliance such as a gas furnace?

SOLUTION

The same *chimney*, sometimes. The same *flue*, never! A single chimney can have multiple flues (insulated, fireproof air ducts leading to the outside), each of which serves a different appliance. If your house has a chimney, you should have it inspected by the fire marshal, and by your insurance company, before you use any appliance that vents into it.

TIP *If a single chimney has more than one flue, you can safely vent one, but only one, home heating appliance into each flue.*

WARNING! *Every home heating system should vent into its own dedicated (separate) flue or exhaust outlet. You'll create a dangerous situation if you connect a wood or pellet burning system to the same flue as any other appliance.*

Corn Stoves and Furnaces

As an alternative to wood pellets, you can burn shelled dry corn in *corn stoves* and *corn furnaces*. These systems resemble pellet stoves and furnaces in many ways, but some important differences exist.

How They Work

Figure 1-4 can serve as a simplified functional diagram of a corn stove. The most obvious difference between the corn stove and the pellet stove is the fact that, rather than wood pellets, individual corn kernels (cleaned of the cob, corn silk, and other foreign matter) go into the hopper. The feed system supplies the kernels to the firebox.

Corn contains significant amounts of *ethanol* (the same form of alcohol used in alternative fuels, such as *gasohol* or *E85* for cars and trucks), whereas wood contains relatively little. Ethanol burns hotter than wood does. In addition, corn contains oil (the same stuff you can use to fry food), which also burns hotter than wood, although more slowly than ethanol.

The waste matter in a corn stove accumulates in the form of a *clinker*, which resembles a concentrated lump of coal. You must periodically remove and discard the clinker, just as you must get rid of the ash from a pellet system from time to time.

TIP *Corn stoves are designed to burn only dry corn, refined especially for use as fuel. Don't try to burn fresh corn, popcorn, corn on the cob, or any other form of corn in one of these things!*

TIP *Corn stoves are rated from approximately 30,000 to 70,000 Btu/h. Corn furnaces, designed for the forced-air heating of entire homes, can deliver considerably more heating power, in some cases upwards of 100,000 Btu/h.*

Advantages of Corn Stoves and Furnaces

- Corn stoves run more efficiently than cut-wood stoves do. The corn kernels are dry and clean, and burn almost completely. As a result, you get maximum heat with minimum pollution and waste matter.

- Corn stoves present less of a danger than cut-wood stoves do, for the same reasons that apply to pellet stoves.

- With a corn heating system, you can easily regulate the temperature, just as you can do with a pellet system.

- You can dispose of the clinker without extended downtime, and it makes less of a mess than the ash that results from the burning of cut wood or wood pellets.

- Corn stoves and furnaces, like pellet systems, don't require chimneys, so you don't have to worry about the problems associated with chimneys. The exhaust gases can vent out through an exterior wall.

- Corn-based heating systems may be allowed in regions or municipalities where wood-burning systems are forbidden.

- If you carry on a friendly relationship with a farmer who produces a surplus of corn almost every year, you might be able to get fuel for free if you're willing to process and dry it yourself.

Limitations of Corn Stoves and Furnaces

- If the electrical power fails, a corn stove won't work. It's the same problem that occurs with a pellet system. You can get a backup battery to run the auger and the blower, but you have to keep the battery charged. It will work for only a few hours in the absence of AC power.

- Corn stoves and furnaces, like their pellet counterparts, contain electronic circuits that can malfunction when power-line "spikes" occur. A corn-burning system should, therefore, employ a transient suppressor in the AC power line.

- If a foreign object gets into the feed system, you'll have the same trouble as you'll have if it happens in a pellet-burning stove or furnace.

- Refined, dried corn, while easily available in some locations, is impossible to get in other places, unless you have it shipped to your location at great expense.

⬛ **PROBLEM 1-3** _____

Do the kernels in a corn stove or furnace ever pop uncontrollably, causing noise and a mess, and giving rise to the risk of an explosion?

✅ **SOLUTION** _____

Corn kernels won't burst in a properly operating system. In fact, the corn won't even snap or crackle. If the above mentioned scenarios were a problem, corn stoves wouldn't have survived on the market (except perhaps as novelty devices).

Coal Stoves

The United States has abundant coal reserves. Because of this fact, and also on account of increasing prices for conventional heating fuels and the continued exotic nature of more technologically advanced heating methods, coal-burning stoves have become popular in recent years.

How They Work

Coal stoves resemble wood stoves. In fact, hybrid units exist that can burn either coal or wood. The main difference between a coal-burning stove and a wood-burning stove lies in the nature of the air intake system. Wood burns best with air supplied mainly from above, while coal burns better when the air comes in from underneath.

Figure 1-5 is a functional diagram of a hybrid stove that can burn either coal or wood. For wood burning, you close the coal air intake damper, and you use the wood air intake to adjust the air flow and the rate of combustion. For coal burning, you close the wood air intake and open up the coal air intake and damper. In either situation, the fuel produces ash that collects in a pan at the bottom of the stove. You must periodically remove and empty this pan.

Some coal-burning stoves have thermostat-controlled dampers that regulate the air flow, and consequently, the burn rate. A more advanced coal-burning stove may have a blower at the coal air intake point to increase the circulation and improve the efficiency. Such a blower requires electricity, so it won't work in the event of a power outage (unless you have a backup power source such as a battery or generator).

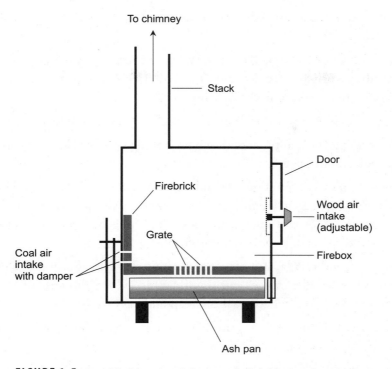

FIGURE 1-5 • Simplified functional diagram of a hybrid stove that can burn cut wood or anthracite coal.

TIP   *Coal-burning stove manufacturers recommend the burning of* **deep-mined anthracite** *coal. They generally discourage the use of* **bituminous** *coal.* **Peat** *and* **lignite** *will burn in some hybrid systems just as wood does, but you should consult the stove's instruction manual before you make that substitution.*

TIP   *You should never use a wood-only stove in an attempt to burn coal. The air intake system won't work properly for any fuel other than dry, cut wood. Similarly, you shouldn't try to use a coal-only stove to burn anything other than deep-mined anthracite. If you want hybrid performance, you need a hybrid stove!*

Advantages of Coal Stoves

- Coal is easily available and is also economical in some locations, making coal a viable alternative energy source for people who live in those places. (If you're one of these people, you know it!)

- Anthracite coal burns fairly clean, contrary to the widely held belief that all coal is "dirty."
- Coal stoves offer high efficiency. When properly installed with an air distribution system, a coal stove can heat a small house to a comfortable temperature even in frigid weather.
- Coal does not have to be manufactured, as do wood pellets.
- The availability and price of coal does not depend on what happens in the agricultural market, as is the case with corn.
- Some coal stoves function without augers or other electromechanical feed systems, getting rid of the problems that go along with those devices. However, other systems do include electromechanical feed systems to meet the needs of people who prefer them.
- Frequent and regular use of a coal-burning stove, in addition to a conventional gas or oil furnace, can significantly reduce the cost of heating a home in locations where coal is cheap and easy to get.
- A coal stove can serve as an emergency heat system in the event of an interruption in conventional utility supplies.

Limitations of Coal Stoves

- Coal stoves can pose a physical danger to life and property if not operated correctly, for the same reasons wood stoves can be hazardous.
- You must maintain a supply of coal at the home site, presenting an inconvenience. In addition, some people don't like the idea of having a "coal pile" around.
- Although coal doesn't pollute as much as some people imagine, it's not as clean-burning as oil or natural gas.
- A coal fire requires a lot of attention, unless the stove can hold a large quantity of coal and has an automatic feed system.
- A coal stove requires frequent cleaning. The chimney also needs periodic cleaning and inspection to ensure that it remains in good overall condition.
- If you want to heat your whole house with a coal stove, the room where the stove is located will become extremely hot unless you have an air-distribution system.

- People in some locations will find coal difficult or impossible to get at a reasonable price.
- Some local or municipal governments restrict or prohibit the use of coal-burning stoves.

? Still Struggling

Suppose that you want to install a wood, pellet, corn, or coal stove to back up your "main furnace," which burns natural gas or oil. Before you rely on an alternative heating source of this sort to serve a major role in heating your home or business, consult the people in your local fire department. Consult the fire marshal. Pick their brains! Obtain all the pamphlets and other data you can from them, heed all local regulations, make sure that your insurance company knows (and approves of) what you're doing, and get a final, official inspection of the system after installation. Have fire extinguishers handy, use smoke detectors and a *carbon monoxide* (CO) detector in your house, and devise an emergency evacuation plan. Some towns and counties offer fire-safety seminars and classes. Take advantage of them.

PROBLEM 1-4

I've read stories about the use of coal and wood stoves during severe winters of the Northern Great Plains in pioneer days. One story told of how the stove got so hot that it glowed. Isn't this dangerous?

SOLUTION

The short answer is yes! If a coal or wood stove gets so hot that it visibly glows, the combustion rate is too high. This state of affairs, if allowed to continue for a long time, can permanently ruin the stove, and might cause sudden structural failure of the firebox. Then you'll have red-hot coals all over your floor! If you think a coal or wood stove is getting too hot, turn it down by restricting the air intake.

QUIZ

Refer to the text in this chapter if necessary. A good score is eight correct. You'll find the correct answers listed in the back of the book.

1. Suppose that a pellet stove can produce 40,000 Btu/h of useful heat power output. That's the equivalent of
   A. 37.91 kW.
   B. 136.5 W.
   C. 11.72 kW.
   D. 2638 W.

2. It takes exactly 1 cal of heat energy to
   A. warm up 1 g of pure liquid water by 1°C.
   B. melt 1 g of pure water ice entirely into liquid.
   C. completely vaporize 1 g of pure liquid water.
   D. completely vaporize 1 g of pure water ice.

3. We can express the rate of energy expenditure at any particular moment in time, or averaged over a period of time, in
   A. British thermal units.
   B. joules.
   C. calories.
   D. kilowatts.

4. A corn stove will best burn
   A. popcorn.
   B. shelled dry corn.
   C. corn on the cob.
   D. Any of the above

5. In order to protect the sensitive electronics in a pellet stove, you should always use
   A. a rechargeable backup battery.
   B. a transient suppressor in the power line.
   C. a dedicated flue.
   D. an electric air blower.

6. If you want a backup heating stove that will work in case the electric utilities go down, and that doesn't need a battery or generator, you'd most likely choose a stove designed to burn
   A. cut wood.
   B. wood pellets.
   C. corn.
   D. Any of the above

7. **A gas furnace and a wood stove can safely vent into the same chimney**
    A. only if the gas furnace vents into the same flue as the wood stove does.
    B. only if the gas furnace vents into a different flue from the one the wood stove uses.
    C. only if both systems always run simultaneously.
    D. under no circumstances.

8. **We can measure accumulated power consumption over a period of time in**
    A. watts.
    B. British thermal units per hour.
    C. joules.
    D. kilowatts per hour.

9. **Which of the following devices is designed specifically to reduce particulate pollution from a wood stove?**
    A. A dedicated (separate) flue
    B. An air blower
    C. An adjustable air intake
    D. A catalytic converter

10. **Which of the following fuel types burns best in a coal stove?**
    A. Deep-mined anthracite
    B. Lignite
    C. Peat
    D. Bituminous

chapter 2

Heating with Oil and Gas

Centralized heating systems transfer thermal energy by circulating air, water, or steam, heated with devices fueled by oil or gas extracted from deep within the earth. (Some people call these flammable substances *fossil fuels*.) Before we examine the three major types of fossil fuel used for home heating, let's compare the most common heat distribution methods.

CHAPTER OBJECTIVES

In this chapter, you will

- Learn how forced-air central heating systems operate.
- Compare hot-water and steam heating systems.
- Learn how embedded radiant heating and radiant-heat subflooring work.
- Contrast oil, methane, and propane as heating fuels.
- Learn how a forced-air fan can function with a wood, corn, or coal stove to heat your entire house.

Forced-Air Heating

Forced-air heating can function in systems that operate from any fuel source. In this type of system, a central furnace heats the air, which circulates throughout the house with the help of a large blower and a network of intakes, ducts, and vents.

How It Works

Figure 2-1 is a functional diagram of a forced-air heating system. In this example, the *circulation air intake* is located inside the building. Some systems take in air for circulation from the outside instead. In contrast to the circulation air

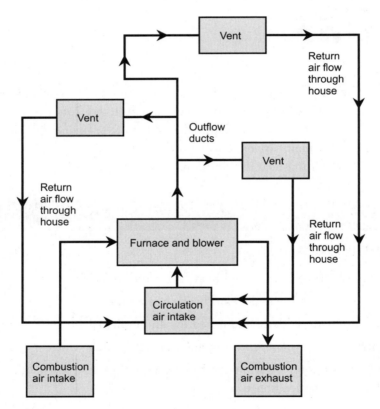

FIGURE 2-1 · Functional diagram of heat distribution in a forced-air gas- or oil-fired furnace using ducts and vents. Arrows indicate the direction of air flow.

intake, a second intake called the *combustion air intake* is almost always located outside. This air provides the oxygen for the fuel-burning process only, not for the house.

TIP *If a forced-air heating system's combustion intake is located indoors, the extra air flowing into the furnace will create negative pressure inside the house, causing cold outdoor air to enter the structure wherever a leak exists. That effect will reduce, not improve, the overall efficiency of the heating system.*

Air from the circulation intake flows into the furnace. The *firebox* heats this incoming air. A *furnace blower*, also called a *fan*, pushes the heated air into a network of *ducts*, where it travels to *vents* (also called *registers*) in the rooms. A small room typically has one vent, but large rooms may have two or more. The vents are located at the base of a wall, or in the floor near a wall. As long as the vents remain unobstructed, they distribute warm air efficiently throughout the room by convection.

The air circulation intake, if located indoors, draws air back to the furnace from inside the house, and also to a small extent from the outside, because no house is (nor should be) perfectly airtight. If the circulation intake is located outside, none of the warmed air recirculates. An outdoor circulation intake reduces the risk and severity of deadly *carbon-monoxide* (CO) gas buildup in the house if something goes wrong with the system.

TIP *The exhaust from a forced-air furnace, which contains noxious gases, vents to the outside through a chimney or side vent. You should exercise care to ensure that the exhaust vent always remains unobstructed. In the winter, you should frequently inspect the exhaust vent and make certain that it's never blocked by snow or ice. Obstruction of the exhaust vent can result in CO gas accumulation indoors, even if fresh air enters through an outdoor air circulation intake. Modern high-efficiency furnaces are designed to shut down in the event of exhaust-vent blockage.*

Advantages of Forced-Air Heating

- If you leave the house for a few days, you can turn the thermostat down to a low setting (just warm enough so the water pipes won't freeze), and when you return, set it back up again, and the house will rapidly rewarm.

- You can set the thermostat low at night while you sleep, and set it higher during the daytime.

- You can operate a forced-air heating system in conjunction with a humidifier in dry climates, minimizing problems with electrostatic charge buildup ("static electricity") and maintaining a healthy indoor environment.

- In damp climates, a forced-air heating system, used without a humidifier, tends to dry the air somewhat, discouraging condensation (particularly in cool basements) and the growth of mold.

- In a forced-air system with an indoor air intake, you can add a cleaning filter or air ionizer to remove particulates and allergens from the air.

- Forced-air furnace ductwork can serve as air conditioning (cooling) ductwork during the warmer months.

- If you have a forced-air heating system, you can use its fan in conjunction with a wood, corn, or coal stove to heat your whole house in case of an interruption in the normal fuel supply. Such an alternative system, if located near the blower intake, can also "help out the main furnace" during extremely cold weather.

Limitations of Forced-Air Heating

- A forced-air system that draws air from a dusty outdoor environment will introduce dust into the home. (Some people in desert locations use the term "forced dust" to describe the effect!) Air filters can get rid of some, but not all, of this dust. This problem can be mitigated by using an indoor air circulation intake rather than an outdoor one. It also helps to set the fan to "automatic" mode so it runs only when the furnace generates heat.

- You must replace the air filters in the furnace frequently, or the entire system will become inefficient (and ultimately break down) because it will have to work harder than it should to circulate the air.

- If a critical component fails, you'll have no heat. You should consider getting a service contract with a vendor whom you know will always be available and have parts for your particular furnace. (Trust me on this issue. I speak from experience.)

- In the event of any malfunction that causes CO gas to enter the indoor circulation, that gas will rapidly distribute throughout the house, posing a danger to all the occupants (humans and pets).

PROBLEM 2-1

Why should a house not be completely airtight? Wouldn't a forced-air system that draws air from the inside, in a completely airtight house, offer optimum energy efficiency and, therefore, find favor?

✔ SOLUTION

An airtight house has better energy efficiency (if all other factors hold constant) than a house with significant air leakage. But problems can occur in houses that are too airtight. If a furnace malfunction introduces CO gas into the circulated air, the CO may reach deadly levels before you can react, even if you have installed CO sensors with alarms. The danger multiplies if such an event occurs while you sleep.

Hot-Water and Steam Heating

Updated versions of "old fashioned" hot-water and steam heating systems have gained popularity in recent years, especially *embedded radiant heating* and *radiant heat subflooring*.

How They Work

In theory, any technology that can get water to boiling or near-boiling can operate a hot-water or steam heating system. Methane, oil, and propane heaters work well for this purpose. Some older buildings used coal-burning systems to fire their boilers before they were converted to oil. Alternative fuels, such as wood or corn, can also operate water heaters or boilers.

Figure 2-2 is a functional diagram of a system in which hot water or steam flows throughout the house, following a network of pipes. In the rooms, the hot water or steam passes through complex metal structures having minimum internal volume and maximum exposed surface area. This design optimizes the transfer of heat energy from the water or steam to the surrounding environment. If the structures are directly exposed to the air, we call them *radiators*. If they're embedded in the floor and/or walls, we call them *coils*.

In a steam system, after heat energy has gone into the air by way of the radiators, the vapor condenses and returns to the boiler as hot water. The boiler heats the water until it changes to the vapor state once more, and sends the steam on

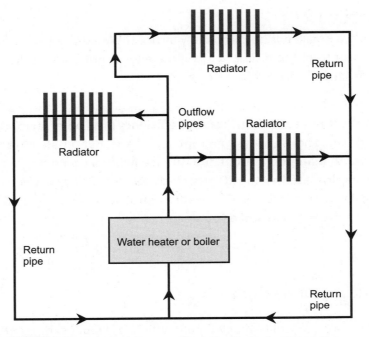

FIGURE 2-2 · Functional diagram of heat distribution in a hot-water or steam furnace using pipes and radiators. Arrows indicate the direction of water or steam flow.

its way through the house again. In a hot-water system, the water returns to the water heater at a lower temperature than that at which it left, and emerges from the heater ready for another round.

Advantages of Hot-Water and Steam Heating

- Hot-water heating plants constitute *closed systems* in theory. A perfect system wouldn't consume any water after the initial charging. (In practice we can approach, but not quite attain, this ideal.) Closed systems work efficiently!

- Because no fan exists, positive or negative pressures don't build up. This feature minimizes the amount of energy wasted in "heating the outdoors" if warm air escapes and/or cold air enters because of pressure differences between the inside and the outside.

- Hot water and steam heating systems don't introduce dust into, or distribute dust throughout, a building.

- Deadly CO gas can't circulate very fast throughout a building in the event of a malfunction that causes the heating unit to emit this gas. However, you must place a CO detector near the heating unit to provide advance warning if a problem does occur.

- Radiant heat coils embedded in the floor or walls don't intrude into rooms, and remain invisible.

- Hot water *baseboard radiators* have a low profile, although you'll have to keep combustible materials away from them. Some clearance should also be maintained, so the radiators can transfer heat adequately into rooms.

- If you have radiant heat subflooring, you can wake up to a warm hardwood or laminate floor, even on the coldest winter mornings.

Limitations of Hot-Water and Steam Heating

- It takes a long time to completely heat up a cold house by means of warm embedded objects.

- Hot water can leak if radiators, coils, or pipes rupture, rust, or fracture. Such a mishap can cause water damage to surrounding objects and structures.

- Steam radiators will leak if the pipes rupture, rust, or fracture. This sort of event can damage nearby objects, and can cause serious burns to people. You'll know immediately if a steam radiator springs a leak; the hissing or squealing noise will give it away!

- Older steam radiator systems are notorious for banging and clanging as the pipes expand and contract. These noises don't normally occur in well-designed, properly operating, newer hot-water systems.

- Any type of radiator becomes hot to the touch and can burn a person who comes into direct contact with it.

- In a hot-water system, you must keep the water free of minerals to prevent deposits from building up in the pipes, water heater, and/or boiler. For that reason, you'll have to install a water softener if your house doesn't already have one.

 **PROBLEM 2-2**

Can radiant heat subflooring function properly with carpeted floors?

☑ **SOLUTION**

Yes, provided that you install the carpet with matting underneath that does not provide too much thermal insulation. You'll need to employ the expertise of a professional installer to get the best results.

Oilheat Technology

Many homes rely on oil for heating, particularly in the northeastern United States. *Oilheat technology* has undergone a renaissance in recent years. Engineers have developed high-efficiency, clean-burning oilheat systems that compete favorably with systems that use other types of fuel.

How It Works

An oil-fired central heating system comprises several components. A *fuel tank* is located on the property. This tank can sit above ground or lie below ground, depending on the location, ordinances, and covenants in the neighborhood, and the preference of the property owner. A *pipeline* runs from the tank to the furnace unit. Service personnel periodically fill the tank.

The furnace *atomizes* (renders as a fine mist) the heating oil in a manner similar to the way a carburetor atomizes the gasoline in a motor vehicle. The process involves breaking the oil into fine particles that mix with air. An electric igniter burns the oil/air mixture. The resulting flames release heat in a *combustion chamber*. This heat can supply a forced-air distribution system, a water heater and radiator system, or a steam boiler and radiator system.

In older homes, steam radiator systems commonly serve in conjunction with oilheat furnaces. In newer homes with oilheat technology, radiant heat subflooring, in conjunction with a water heater, is popular. Some homes employ hot water baseboard radiators comprising pipes with steel plates attached at right angles, maximizing the surface area while minimizing the volume (Fig. 2-3). These radiators run along the floor at the base of a wall.

The main by-products of oil combustion are water vapor (H_2O) and carbon dioxide (CO_2) gas. Trace amounts of sulfur dioxide (SO_2) gas, CO gas, and particulate matter also form. All of these waste materials vent to the outside as exhaust, either through a conventional chimney flue, or with high-efficiency oilheat systems, through a plastic *flue pipe* passing through an outer wall of the house.

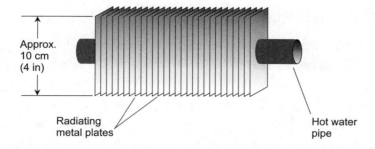

Approx.
10 cm
(4 in)

Radiating
metal plates

Hot water
pipe

FIGURE 2-3 · Simplified diagram of a hot-water baseboard radiator section. In a typical system, vents cover the elements.

TIP *You should always cover steam or hot-water baseboard radiators with metal enclosures (also known as registers) to protect the radiator elements against dust accumulation, and to prevent physical damage to the plates if the assembly is bumped. The registers also reduce the risk of burns to people who accidentally come into contact with them.*

Advantages of Oil Heating

- An oilheat system has no pipeline to a central supplier. You can (and usually must) locate the tank on your own property. This factor holds significance for people who want to live in rural or remote areas, far from urban centers.

- Oil is a relatively safe fuel. An oil leak can produce a mess, but it doesn't pose the risk of an explosion as gases do. Oil is less explosive than any flammable gas, and is also less explosive than gasoline, alcohol, or kerosene.

- If an oil-fired system malfunctions, smoke almost always comes out. You can't miss it! The stench calls attention to the problem.

- Oil is a high-density fuel. That means a small volume and mass of oil contains a large amount of *potential energy*. A modest-sized, on-site oil tank can provide several weeks' worth of central heat for a household, even in bitter-cold locations.

- Because of its high density, oil is a portable fuel. Small oilheat systems work well in recreational vehicles, hunting cabins, or other small, temporary living spaces.

- Heating oil can mix with *biofuels*, such as surplus vegetable oil, used cooking grease, or discarded animal fat. Specially designed oilheat systems can efficiently burn these so-called *hybrid fuels*.

Limitations of Oil Heating

- You must carefully watch the on-site storage tank so that it won't go empty at any time during the cold season. If it does, you'd better have a backup heating system ready to take over until the oil supplier can come out!
- The price of heating oil more or less follows the price of crude oil. This price can spike rapidly. Pundits will endlessly debate the price prognosis for crude, but we should all expect it to rise over the long term.
- Much of the world's crude oil comes from countries with a history of political instability. As a result, the risk always exists for a sudden reduction in the supply, with skyrocketing prices and perhaps critical shortages as a result.
- Temporary reductions in the crude oil supply (for the United States, at least) can be caused by massive hurricanes in the Gulf of Mexico.
- The world has a finite supply of crude oil, and it's not renewable. (In contrast, biofuels, mentioned above, are renewable.)
- Certain components in modern oilheat systems require electricity. If you want the furnace to work during an electrical power failure, you need a backup power source that must be ready to go at any time. It must be capable of delivering a clean sine wave (so the electronic control circuits in the heating system will work), and it must be able to provide enough current to operate the whole furnace including all its components.

TIP   *Consult a professional electrician concerning the installation of a backup generator for use with any heating system.*

PROBLEM 2-3

Won't oil freeze or grow sluggish in cold weather, preventing free flow from the tank to the furnace?

✔ SOLUTION

Heating oil won't actually freeze at temperatures encountered in any inhabited region on this planet. Heating oil can become sluggish if it gets cold, just as motor oil does, but you can protect the pipes in an oil-heat system in the same way as you protect water pipes from freezing.

If possible, you should place your residential oil tank underground, where the temperature does not vary greatly even when exceptional "cold snaps" occur in midwinter.

Methane ("Natural Gas") Heating

During the twentieth century, *methane* (CH_4), often called "natural gas" or simply "gas," became the most popular fuel for central heating in the United States. (Actually, methane is one of several constituents of *true natural gas* that we extract directly from the earth.) During the early twenty-first century, periodic methane shortages began to occur. Methane nevertheless remains a viable heating fuel choice, particularly in urban centers where gas pipeline systems already exist.

How It Works

The geometric configurations of gas burning furnaces vary greatly, but they all operate according to the same basic principles. Let's take a simplified look at the operation of a conventional forced-air gas furnace.

The *thermostat* determines when heat is called for. It comprises a simple switch that starts the process of furnace operation. You'll usually find it in the main living space or a major hallway. The *induced draft motor*, also known as the *combustion air blower* or the *inducer*, supplies the air necessary to burn the fuel, and also blows away the exhaust gases resulting from combustion. These gases can vent out through a chimney flue. In a high-efficiency furnace, these gases are *condensed* to extract additional heat, and the cooled exhaust vents out of the house through a plastic flue pipe.

The *inducer* starts running as soon as the thermostat calls for heat. A *centrifugal switch*, also called a *pressure switch*, verifies that the inducer is operating. (If either the inducer or the pressure switch fail, the furnace will not produce any heat.) After a short delay, the *gas valve* opens, and an electric igniter, also known as a *glow plug*, sets the fuel on fire. In older systems, a *pilot light* facilitates the combustion, and a temperature-actuated electronic switch called a *thermocouple* keeps the gas valve from opening if the pilot light has gone out. *Burners* in the *firebox* provide the combustion that warms up the *heat exchanger*.

In a forced-air system, the furnace fan (furnace blower) circulates the warmed air from the heat exchanger throughout the house. The blower starts after a delay period of 1 to 3 minutes following burner ignition. When the temperature in the house rises to a certain level, the gas valve closes, and the burners go off. Then the inducer stops running. After a delay of 1 to 3 minutes, the fan shuts down if the system is set for *automatic (auto) fan mode*.

TIP *In most forced-air gas heating systems, the operator has the option of leaving the fan on constantly, so air keeps circulating even when the furnace produces no heat. This mode can minimize thermal stratification, where cold air tends to accumulate in certain rooms, or near the floors of all rooms. If you have a wood, corn, or coal burning stove located near the gas furnace air intake, and if you leave the furnace fan running continuously, you can use the stove to heat the house as an alternative, or as a supplement, to the furnace.*

TIP *Some gas furnaces operate with water heaters or boilers. Some systems of this type employ multiple thermostats to regulate the temperature in individual rooms or portions of the house, a scheme called* zone heating. *The thermostats actuate valves that control the flow of water to the various radiators or coils.*

Advantages of Methane Heating

- Methane heating systems can operate with exceptional efficiency. Modern furnaces with exhaust condensers can convert nearly all of the potential energy in the methane into usable heat.

- High-efficiency gas furnaces can vent exhaust directly through a wall to the outside, eliminating the need for a chimney. If you install a high-efficiency gas furnace in place of an oilheat system or an older gas furnace, the chimney flue for the displaced system can serve a supplemental heat source such as a wood, corn, or coal stove, as long as you don't use that flue to vent any other heating appliance.

- Natural gas is readily available in most cities and towns. Underground pipelines provide an uninterrupted supply. You don't have to worry about the status of an on-site supply tank.

- Gas furnaces (as well as gas stoves, and fireplaces, and water heaters) burn clean. They produce relatively little air pollution, and essentially no smoke.

- You can obtain a methane heating system in an *upflow*, *downflow*, or *horizontal* (or *lateral flow*) design. This flexibility allows you to install the main furnace unit in the basement, a modular or mobile home, the attic, or a crawl space.

Limitations of Methane Heating

- Methane leaks, like any flammable gas leak, can cause flash fires or explosions. In most locales, methane gas has an artificial scent that you can easily recognize, and that alerts people to the existence of gas leaks. If you "smell gas" in your house, *get out of there immediately* and call the fire department.

- In recent years, the price volatility, and occasional problems with the supply, of natural gas has injured its reputation as the most reliable home heating fuel.

- In rural locations, and in less developed parts of the world, permanent gas pipelines don't exist.

- Methane gas is not easy to store for a single household, unless you own a large farm or ranch and the local laws or covenants allow methane storage tanks on private property.

- The world's supply of naturally occurring methane is finite and nonrenewable. (However, certain biological processes can produce methane artificially, providing a renewable source of the gas.)

- Certain components in modern methane heating systems require electricity. They will not work if the utility electricity fails, unless you have a backup power source. See the last limitation note for oilheat systems, above.

PROBLEM 2-4

Can other gases replace methane in a heating system designed to burn methane? How about hydrogen, in particular?

✔ SOLUTION

Hydrogen holds some promise for eventual use in place of methane as the piped-in fuel in cities and towns. Its outstanding advantage is the fact that it doesn't produce any pollution; when it burns, all you get is heat and water vapor! But hydrogen, being the lightest chemical element, leaks more easily than methane. In addition, it burns hotter. The higher operating temperatures will necessitate changes in pipeline and furnace design. As of this writing, even though hydrogen constitutes the most abundant element in the universe, it does not occur in its free form on earth in quantities that make it easy to utilize as fuel.

TIP *Methods exist to separate and isolate hydrogen from chemical compounds containing it. For example, direct electric current through salt water will split the water molecules into atoms of hydrogen and oxygen; the gases can be collected and stored. However, at the time of this writing, all existing hydrogen-production technologies remain expensive and inefficient.*

Propane Heating

The term *propane* usually applies to *liquefied petroleum* (LP) gas. In the United States, LP gas consists mainly of the hydrocarbon technically known as propane (C_3H_8), hence the name. In some countries, suppliers deliberately add significant quantities of *butane* (C_4H_{10}) to the propane in LP gas. Propane, in its gaseous state, weighs more than methane per unit volume, and butane weighs more than propane per unit volume.

How It Works

Propane occurs naturally as a by-product in the process of extracting methane from true natural gas, or as a by-product in the process of refining crude oil. Propane can remain stored under pressure in liquid form for quite a while. When the liquid escapes from the pressurized tank into the atmosphere, it becomes a flammable gas.

Propane has no color and no odor; the same holds true for butane and methane in the gaseous state. Industrial engineers add a compound called *ethyl mercaptan*, which smells something like rotten eggs, to propane to facilitate easy detection of leaks. Once you've smelled propane gas, you'll never forget it (unless you happen to be one of the few people whose noses are "numb" to it).

The fact that propane can be stored in tanks makes it portable. For this reason, propane serves as a popular alternative fuel for home backup generators, generators for use with recreational vehicles, and portable cooking stoves. Propane can burn in a central home heating system as an alternative to oil or methane. It can also fuel motor vehicles.

TIP   *When propane depressurizes as it escapes from a tank, it burns in much the same way as methane gas does. In fact, propane and methane furnaces have nearly identical designs.*

Advantages of Propane Heating

- You can use propane almost anywhere because it stores well in tanks. When you see a white or silver tank shaped like a fat sausage in the yard of a rural home, farm house, or ranch house, you can have reasonable certainty that you're looking at a propane tank.

- Propane does not dissolve in water, as many other flammable materials (including gases) do. For this reason, propane fuel presents little risk of contamination to ground water or soil.

- Propane heating systems pollute less than wood or coal burning systems. Propane systems compare favorably to oil and methane systems in terms of the amount of pollution they cause per unit of energy produced.

- As an alternative fuel, propane can sometimes offer cost savings to people who live in areas where methane and oil prices are high.

- With a propane system, you can meet all the energy needs of your home. With a propane generator, you can live disconnected from the electric utility grid (although with considerable sacrifice).

- Propane does not burn as rapidly as methane does, so propane poses less of an explosion hazard. However, you should take the same precautions with a propane system as you would do with a methane system.

Limitations of Propane Heating

- The price of propane may not be the lowest price for any fuel available in a given location. When considering price, you must include all factors, such as the cost of furnace conversion or installation (if required) and maintenance, the risk of interruptions in the fuel supply, and the attitudes of homeowner's insurance companies (reflected by their rates) toward the use of various fuels in a particular location.

- Propane, while stored in liquefied form, does not exhibit the highest energy content per volume of any conventional fuel choice. That distinction goes to home heating oil. Because of this fact, propane storage tanks must have greater volume, per unit of energy delivered, than oil storage tanks.

- If the tank temperature falls to extremely low levels, as it does in some northern parts of the United States, much of northern Europe, and much of Canada, propane may not revert to the gaseous phase when released. This will cause the heating apparatus to become inoperative.

- Not everyone can detect the odor of ethyl mercaptan in LP gas.

- If oxidation (rust) occurs inside the tank or pipeline, *odor fade* may occur. In that case, you might not detect a propane leak based on your sense of smell alone. *Propane gas detectors* are recommended in any propane installation for this reason.

- Certain components in modern propane heating systems require electricity. They will not work if the electrical power fails, unless you have a backup power source. See the last limitation note for oilheat systems, above.

? Still Struggling

All of the fuels discussed so far involve "burning something." You might wonder when we're going to get away from the combustion concept! When any flammable material except hydrogen burns, air pollution results (sometimes a lot, sometimes a little, but always *some*). Most of these fuels come out of holes that we drill in the earth. You ask, "Why can't we humans free ourselves from this ancient business, and seriously consider the long-term well-being of our planet?" That mantra plays well with environmental groups. Most scientists agree that the old ways must pass or humanity faces potentially devastating environmental and climatic consequences. But "archaic" energy technologies will remain the most cost-effective home heating alternatives for much of the world's population for decades to come. Transition we must—but it will take time.

QUIZ

Refer to the text in this chapter if necessary. A good score is eight correct. You'll find the correct answers listed in the back of the book.

1. **In a forced-air heating system, how does an indoor circulation intake compare with an outdoor circulation intake, if all other factors hold constant?**
    A. An indoor circulation intake facilitates better efficiency than an outdoor circulation intake does.
    B. An outdoor air intake can freshen the air in the house more than an indoor air intake can, assuming that the outdoor air actually is fresh.
    C. An indoor air intake increases the risk or severity of CO gas buildup in the event of a system malfunction, compared with an outdoor air intake.
    D. All of the above

2. **With a central forced-air heating system, you can reduce the buildup of "static electricity" in dry environments by adding**
    A. a centrifugal switch.
    B. an inducer.
    C. a humidifier.
    D. a catalytic converter.

3. **Propane systems present little risk of polluting nearby water wells because propane**
    A. doesn't dissolve in water.
    B. can't soak into the ground.
    C. never leaks out of tanks.
    D. can't escape into the air.

4. **Hydrogen combustion produces**
    A. carbon compounds.
    B. methane gas.
    C. water vapor.
    D. particulate pollution.

5. **In some places, LP gas contains significant quantities of**
    A. oxygen.
    B. hydrogen.
    C. methane.
    D. butane.

6. **If a steam radiator leaks, you'll know it right away because**
    A. you'll smell the ethyl mercaptan.
    B. you'll sense the decrease in humidity.
    C. you'll hear the noise.
    D. All of the above

7. **In a steam heating system, what happens to the steam after its excess heat energy has gone into the air by way of the radiators?**
   A. It escapes through a dedicated chimney flue or exhaust vent.
   B. It condenses and returns to the central furnace as water.
   C. It remains in the vapor state and returns to the central furnace for reheating.
   D. It condenses and flows into the home's main water supply for general use.

8. **Oil is generally safer than methane as a heating fuel because**
   A. oil is less energy-dense than methane.
   B. oil burns more rapidly than methane.
   C. oil is less explosive than methane.
   D. oil does not freeze the way methane can.

9. **A hot-water heating system does not cause positive or negative pressure buildup inside a house because**
   A. it has no blower.
   B. the air intake and outflow, while both considerable, balance each other out.
   C. the temperature can't change very fast.
   D. excess water vapor equalizes the pressure inside and outside the house.

10. **The device that atomizes the fuel in an oilheat furnace resembles the**
   A. boiler in a steam heating system.
   B. ionizer in a propane heating system.
   C. inducer in a forced-air heating system.
   D. carburetor in a gasoline-powered truck.

chapter 3

Heating and Cooling with Electricity

We can use electricity, directly or indirectly, to heat or cool our homes. Before we learn how such processes work, let's review the ways in which scientists quantify heat energy, and also what can happen when matter warms up or cools down.

CHAPTER OBJECTIVES

In this chapter, you will

- Compare the Celsius, Fahrenheit, and Kelvin temperature scales.
- Convert temperature readings from one scale to another.
- Learn how electric resistance heating systems work.
- Explore the behavior of water when it boils, condenses, freezes, and thaws.
- Discover how refrigerants make electric cooling systems work.
- Learn how electric heat pumps can efficiently warm or cool your house.

Temperature

Temperature quantifies the *kinetic energy* (or heat energy) in a given sample of matter. When we allow energy to move freely from one medium into another as heat, the temperatures of the two media tend to equalize as time passes. Physicists call this process *heat entropy*.

The Celsius Scale

If you place a sample of ice where the temperature is above the freezing point, the ice starts to melt as it accepts heat from the environment. The ice, and the liquid water produced as it melts, has a temperature of 0 degrees on the *Celsius temperature scale* at sea level, a situation that you can symbolize by writing 0°C. As energy flows into the chunk of ice, more and more of the ice melts.

If energy continues to flow into the water once it has become all liquid, its temperature begins to increase. Eventually the water starts to boil, and some of it changes to the gaseous (or vapor) state. The liquid water temperature, and the water vapor that comes immediately off of it, has a temperature of +100 degrees on the Celsius scale (+100°C) at sea level. As energy flows into the water, more and more of it evaporates.

If energy continues to flow into the water once it has all become vapor, its temperature begins to increase again. Ultimately, the only limit to how hot the water vapor can get depends on how much energy the heating elements can deliver into it.

We should note two specific temperature values for pure water: the *freezing point* and the *boiling point*, at which there exist two specific numbers for temperature at normal sea-level atmospheric pressure. Scientists have devised the Celsius temperature scale based on these two points.

TIP   *Some people use the term* centigrade *instead of the term Celsius when referring to the above-described temperature scale. A temperature change of plus or minus one degree Celsius or centigrade (±1°C) represents 1/100 of the difference between the melting temperature of pure water at sea level and the boiling temperature of pure water at sea level. The prefix "centi-" means "1/100," so "centigrade" literally means "graduations by the hundredth part."*

The Kelvin Scale

We can freeze water and keep cooling it down, or boil it all away into vapor and then keep heating it up. Temperatures can plunge far below 0°C, and can rise far above +100°C.

There's an absolute limit to how low the temperature in degrees Celsius can become, but there's no upper limit. We might take extraordinary efforts to cool a chunk of ice down to see how cold we can make it, but we can never chill it down to a temperature any lower than approximately −273.15°C. Scientists call this temperature *absolute zero*. It's the basis for the *Kelvin temperature scale*. Units in this scale are known as *kelvins* (K).

A temperature of −273.15°C equals 0 K. The size of the kelvin increment precisely equals the size of the Celsius increment. Therefore

$$0°C = 273.15 \text{ K}$$

and

$$+100°C = 373.15 \text{ K}$$

On the high end, we can keep heating matter up indefinitely. Temperatures in the cores of stars rise into the millions of kelvins. In the centers of galaxies, quasars, and other extreme celestial objects, perhaps temperatures reach billions (thousand-millions) of kelvins.

TIP *When you want to denote a temperature value in kelvins, you need not (and should not) use a degree symbol. Simply write the number, leave a space, and then write the uppercase K. You don't have to worry about plus or minus signs in the Kelvin scale because Kelvin values can never get negative.*

? Still Struggling

An object at a temperature of absolute zero can't transfer energy to anything else because it possesses no energy to begin with! Scientists doubt that our universe harbors anything that actually has a temperature of absolute zero, although some atoms in the far reaches of intergalactic space come close to that "ultimate chill."

The Fahrenheit Scale

In much of the English-speaking world, and especially in the United States, lay people use the *Fahrenheit temperature scale* (°F). The Fahrenheit increment equals 5/9 of the Celsius increment. In other words, a change of 5°C represents the same increase or decrease in temperature as a change of 9°F does.

The melting temperature of pure water ice at sea level equals +32°F, and the boiling point of pure liquid water at sea level equals +212°F. Therefore, +32°F corresponds to 0°C, and +212°F corresponds to +100°C. Absolute zero on the Fahrenheit scale works out as approximately −459.67°F.

Suppose that you let F represent the temperature in degrees Fahrenheit, and you let C represent the temperature of the same medium in degrees Celsius at the same time and under the same conditions. If you want to convert from degrees Celsius to degrees Fahrenheit, you can use the formula

$$F = (9/5)\,C + 32$$

To convert a reading from degrees Fahrenheit to degrees Celsius, you can use

$$C = (5/9)\,(F - 32)$$

Standard Temperature and Pressure (STP)

Scientists define *standard temperature and pressure* (STP) for experimentation and measurement as follows:

- Standard temperature equals 0°C (32°F), the freezing point or melting point of pure liquid water at standard pressure.
- Standard pressure is the atmospheric pressure that will support a column of mercury 760 millimeters (a little less than 30 inches) tall. It works out to approximately 14.7 pounds per square inch or 101,000 (1.01×10^5) newtons per square meter.

TIP   *We don't think of air as having significant mass because we're immersed in it. But air has a lot of "heft" in large amounts! The density of dry air at STP equals approximately 1.29 kilograms per cubic meter. A parcel of air measuring exactly 4 meters high by 4 meters deep by 4 meters wide, the size of a large bedroom in one of those old mansions with a high ceiling, masses 82.6 kilograms. In the earth's gravitational field, that amount of mass equals the weight of a full-grown man.*

PROBLEM 3-1

What's the Celsius equivalent of 72°F? Round the answer off to the nearest degree.

SOLUTION

Use the above formula for converting Fahrenheit temperatures to Celsius temperatures. Once again, that formula is

$$C = (5/9)(F - 32)$$

In this case, you get

$$C = (5/9)(72 - 32)$$
$$= 5/9 \times 40$$
$$= 22°C$$

PROBLEM 3-2

What's the Kelvin equivalent of a temperature of 80°F? Round the answer off to the nearest whole number.

SOLUTION

First, convert from degrees Fahrenheit to degrees Celsius. When you make the calculations, you get

$$C = (5/9)(80 - 32)$$
$$= 5/9 \times 48$$
$$= 26.67°C$$

You shouldn't round your answer off yet because you have another calculation to perform. The difference between readings in the Celsius and Kelvin scales is 273.15. The Kelvin figure is the greater of the two, so you must add 273.15 to the Celsius reading. If K represents the temperature in kelvins, then

$$K = C + 273.15$$
$$= 26.67 + 273.15$$
$$= 299.82 \text{ K}$$

You can round this off to 300 K. Remember, you should not use a degree symbol when writing a temperature value in kelvins.

TIP *When doing a multiple-step calculation involving approximate values like the one in the solution to Problem 3-2, you should keep some extra digits in the values, and round them off only at the very end of the process. This little trick can minimize the chance that you'll get an inaccurate final result because of* **cumulative rounding errors.**

Electric Resistance Heating

We can obtain heat from electricity if we apply an electrical *voltage* to a resistive element, causing an electrical *current* to flow through that element to make it emit *infrared* (IR) radiation, informally called "heat rays." In some cases, visible and/or *ultraviolet* (UV) rays also come from the element in small amounts, but far less than the amount of IR rays. This method of *electric resistance heating* can work quite well in regions where the winters don't get particularly cold.

Heat, Voltage, and Resistance

Figure 3-1 shows the basic components of an electric resistance heating unit. The *alternating-current* (AC) utility electricity provides the power. The *heating element* forms the heart of the system. It usually consists of a large coil or set of coils made of resistive wire that can withstand high temperatures without melting or breaking.

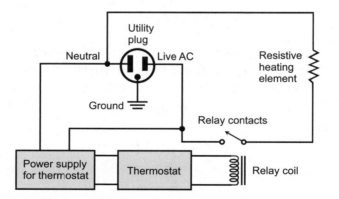

FIGURE 3-1 · Functional diagram of a thermostat-controlled electric resistance heating system designed for use in a single room. In the utility plug symbol, the tall black rectangle represents the neutral terminal, the short black rectangle represents the live AC terminal, and the dot with the multiple-line symbol represents the grounded terminal.

The amount of heat power produced by a resistive element depends on the voltage supplied to it, and also on its resistance. Electricians and scientists express household AC *electrical potential* in terms of *effective voltage*, more often called *root mean square* (RMS) *voltage*. That's the amount of "AC juice" that produces the same amount of heat, if applied to a resistive element, as a *direct current* (DC) voltage with the same numerical value would produce.

If we connect two identical resistive appliances (such as heating elements) to sources of electricity, one AC and the other DC, and if both elements yield the same amount of heat power output as a result of the applied electricity, then the RMS voltage of the AC source (which we can't easily measure) equals the actual voltage of the DC source (which we can easily measure).

If E represents the RMS AC utility voltage in *volts* (V), and if R represents the resistance of the heating element in *ohms*, then we can calculate the power P_W, in *watts* (W), dissipated by the element as

$$P_W = E^2/R$$

TIP   *The typical RMS AC household voltage in the United States is either 117 V or 234 V, plus or minus a few percent. The lower voltage works with small appliances, such as lamps and television sets. The higher voltage works with "power-hungry" appliances, such as electric stoves, ovens, and laundry machines.*

Heat, Current, and Resistance

We can calculate the amount of heat power in watts produced by a resistive element if we know the current I that the element draws from the utility in *amperes* (A), along with the resistance in ohms. In that case, we use the formula

$$P_W = I^2 R$$

This formula and the preceding one involving voltage work, *if and only if*, all the power supplied to the resistive element—every last watt of it—gets converted to heat. In ordinary electric heaters, that always happens "for all intents and purposes," even if the element glows red-hot.

We can modify the foregoing formulas to express heat power in British thermal units per hour (Btu/h). In either case, we multiply the power in watts by 3.41 to obtain the "Btu-per-hour power," $P_{Btu/h}$. As a result, we get

$$P_{Btu/h} = 3.41\ E^2/R$$

and

$$P_{Btu/h} = 3.41\ I^2 R$$

The system portrayed in Figure 3-1 has a *thermostat* along with an electro-mechanical switching device called a *relay*. The thermostat contains a *bimetal strip* that flexes as the temperature rises and falls. This flexing motion opens and closes a set of contacts that supply current to the relay if the temperature falls below a certain point. When the relay coil receives current, its contacts close, and current flows through the heating element. When the temperature rises above a certain point, the bimetal strip flexes back, its contacts open, the relay coil no longer gets any current, and the relay contacts break the circuit so that the resistive element loses its supply of electricity.

Electric Space Heaters

For heating a small room, or for warming up a large room by a few degrees, portable *electric space heaters* are available. Numerous designs exist, all of which employ heating elements with on/off power switches. Some have fans that draw cool air in from the back or the bottom and blow heated air out the front (Fig. 3-2A). Others take advantage of IR radiation from large elements with reflectors behind them (Fig. 3-2B).

Unfortunately, not all electric space heaters have good design. Some of them lack the physical volume and mass to radiate away all the heat power that they generate. As a result, they overheat—not once in awhile, but as a rule! The best electric space heaters have three-wire, heavy-duty cords and exteriors made entirely of metal. Some have plastic casings with metal reflectors inside and a metal grating to prevent users from coming into contact with the heating elements. Some units have elements wound into, or around, ceramic forms. Most (but not all) portable electric space heaters shut themselves off automatically if they overheat or tip over. Let the buyer beware!

WARNING! *Whenever you use an electric space heater, read and heed the instruction manual in every detail. Neglect or improper use can result in burns, fires, or electrocution.*

Radiant Electric Zone Heating

In *radiant electric zone heating*, the elements usually reside in or near the ceiling, minimizing the risk of accidental obstruction, and also minimizing the risk that people will inadvertently touch the hot elements and get burned. Reflectors

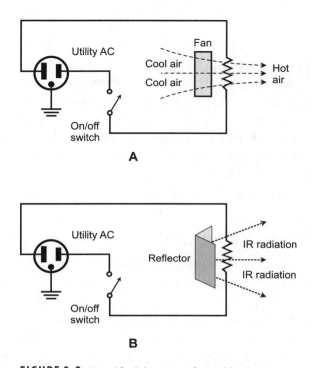

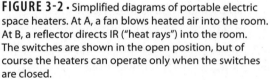

FIGURE 3-2 · Simplified diagrams of portable electric space heaters. At A, a fan blows heated air into the room. At B, a reflector directs IR ("heat rays") into the room. The switches are shown in the open position, but of course the heaters can operate only when the switches are closed.

direct the IR rays downward, and minimize the amount of energy that escapes by conduction through the ceiling or roof.

Thermostats in each room allow the residents to adjust the temperature according to the room occupancy at the time. The radiant system works even though "hot air rises." The primary mode of energy transfer is radiation, which can occur equally well in any direction, even downward. The IR radiation from the ceiling elements warms the people and objects below.

TIP *With a well-designed radiant heating system, you'll feel warmer than the actual air temperature might suggest. The effect resembles the warming effect of the sun's rays on a cool, clear autumn day.*

? Still Struggling

In a radiant heating system, IR energy gets absorbed by solid matter in the rooms, particularly the walls, floors, and furniture. In turn, the warmed surfaces heat the air by convection and conduction. The absorption and secondary warming process takes place most efficiently, and fastest, if the solid surfaces are relatively dark (brown or black, for example).

Baseboard Electric Zone Heating

Wall radiators, located near the floor, facilitate *baseboard electric zone heating*. These radiators resemble the ones found in hot-water or steam heating systems. The same principles apply to electric systems of this kind, as apply to those using other sources of fuel. Electric heating in this application poses no danger of water or steam leakage, obviously. Instead, the danger comes in the form of a fire risk. If a short circuit or lightning strike causes sparks to "fly" from wiring in the radiators, those sparks can ignite flammable materials placed close by.

Electric baseboard radiators are usually installed underneath windows. Heat transfers to the air by convection and conduction. A small gap between the bottom of the radiator and the floor allows cool air to pass upward through the radiating fins and then rise along the wall. Metal enclosures called *registers* surround the fins to protect the fins from damage, and to reduce the chance that people will get burned, or that objects will catch on fire, as a result of accidental contact with the hot fins.

WARNING! *Always allow plenty of clearance between baseboard radiators and objects, such as furniture and draperies. This precaution will minimize the risk of fire, and will also optimize the efficiency with which the radiators can transfer heat energy to the air.*

Central Electric Heating

Some furnaces take advantage of electric resistance principles to provide heat for an entire house. Instead of a firebox where combustion occurs, the electric furnace employs massive resistance elements that require considerable voltage and current to function.

An electric furnace usually operates at 234 V AC. For a given, constant resistance in the heating elements, we get four times as much power from a 234 V AC system as we would derive from a 117 V AC system. Recall the formulas for heat power in terms of voltage and resistance. Once again, they are

$$P_{\mathrm{W}} = E^2/R$$

and

$$P_{\mathrm{Btu/h}} = 3.41\ E^2/R$$

From these formulas, we can see that if we increase the voltage (E) by a factor of 2, then the power (P_{W} or $P_{\mathrm{Btu/h}}$) goes up by a factor of 2^2, or 4. This relation holds valid only if the resistive heating elements can withstand the doubling of current that occurs, can radiate away the fourfold increase in dissipated power, and maintain the same resistance at the higher temperature as compared with the lower temperature.

Electric furnace heating elements are rated at several kilowatts each, and two or more of them are connected together in parallel to obtain the heat for the entire house or building. In effect, the electric furnace operates as an oversized electric space heater as shown in Fig. 3-3.

Advantages of Electric Resistance Heating

- No exhaust gases or waste by-products are created in or around the house.
- If the system malfunctions, dangerous CO gas does not form.
- No explosion hazard exists because the system doesn't involve flammable gases.
- In zone heating systems, individual thermostats can be located in each room, allowing the homeowner to adjust each room's temperature based on real-time use.
- In radiant zone heating systems, dust doesn't circulate throughout the house because no air blower exists.
- Breakers or fuses can provide a cheap, automatic shut-down mechanism in the event of a short circuit, as long as the electrical system has a good earth ground.

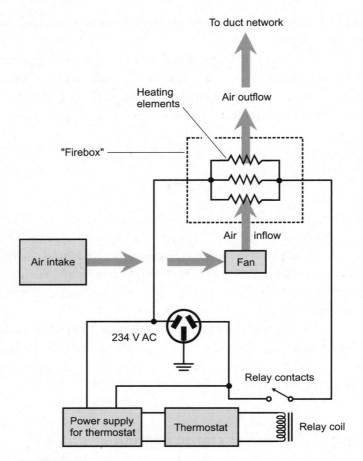

FIGURE 3-3 · Functional diagram of a forced-air electric furnace. Arrows indicate the direction of air flow.

Limitations of Electric Resistance Heating

- Power dissipation by forcing current through a resistance is an inefficient way to heat a room or building, all things considered. All of the electricity gets converted to heat in the unit, but the electric generating plants, from which the power originates, have poorer efficiency than a typical gas- or oil-fired furnace.

- Because of the overall inefficiency of electricity as a heating source, an electric system pollutes the environment more, all stages of the process considered, than a gas- or oil-fired system of equal capacity does.

- Once again considering the relative inefficiency of electric heating, it should come as no surprise that it's an expensive way to heat a home!
- If the utility power fails, electric resistance heating won't work. Onsite backup generators are impractical for ordinary households because of the high current demand.
- Electric space heaters pose a fire hazard if not well made or if improperly used.
- Electrocution can occur if an improperly grounded electric space heater develops a short circuit.
- Electric resistance does not work well as the primary heating mode in households where winters get severe, unless the electric utility rates are exceptionally low.

TIP *In cold-climate regions, some seasonal cabins use electric resistance heating to take the chill out of the indoor air on cool summer evenings.*

PROBLEM 3 - 3

Suppose that a heating unit operates from the standard 117 V AC utility electricity common in American households. If the resistance of the heating element is 18.25 ohms when connected to the electric current source and fully heated up, how much heat power does it produce in watts? How much heat power does it produce in British thermal units per hour? Round off the answers to the nearest whole unit.

SOLUTION

To determine the heat power in watts, use the formula for *wattage* in terms of AC voltage and resistance:

$$P_W = E^2/R$$
$$= 117^2/18.25$$
$$= 750\,W$$

To determine the heat power in British thermal units per hour, multiply the heat power in watts by 3.41 to get

$$P_{Btu/h} = 3.41 \times 750$$
$$= 2558\,Btu/h$$

Principles of Cooling

In descriptions of how cooling systems work, the calorie (cal), defined in Chapter 1, often serves as the energy unit because pure liquid water requires, or releases, exactly one *calorie per gram* (1 cal/g) to warm up or cool down by 1°C. A calorie per gram equals 4.184 *joules per gram* (J/g) or 4184 *joules per kilogram* (J/kg).

Evaporation and Condensation

The above described characteristic, called the *specific heat*, describes water's behavior only as long as the water remains entirely liquid during the heating or cooling process. If liquid water changes to vapor (evaporates), it acquires heat energy, and the surroundings, therefore, lose heat energy, becoming cooler. Conversely, when water vapor changes to liquid (condenses), the water transfers heat to the surroundings, and the surroundings gain heat energy, growing warmer. The same thing happens with other compounds that change state from liquid to vapor or vice-versa.

Suppose that a kettle of water sits on a stove top, and the temperature of the water is exactly at the boiling point (+100°C), but it hasn't started to actively boil. As we apply heat continually, boiling (rapid evaporation) begins. The water becomes proportionately more and more vapor, and less and less liquid. Nevertheless, the temperature of the liquid water in the kettle stays at +100°C. Eventually, all the liquid boils away into vapor. Imagine that we somehow capture all of this vapor in an enclosure along with the kettle, and all the air has been driven out, leaving pure water vapor. Further, suppose that the stove burner keeps on heating the vapor even after it has all boiled out of the kettle! The liquid water couldn't grow any hotter than +100°C, but the vapor can, and it will.

Now let's think about what happens if we take the enclosure, along with the kettle, off of the stove and put it someplace where the temperature is a couple of degrees above freezing. The water vapor begins to grow colder. The vapor temperature eventually drops to +100°C, at which point it starts to condense. The temperature of this liquid equals +100°C. The process goes on until all the vapor in the chamber has condensed. The chamber keeps growing colder. Once all the water vapor has condensed into liquid, the temperature of the liquid water begins to fall below +100°C.

Heat of Vaporization

Interestingly, the temperature of water doesn't rise or fall continuously when heating or cooling takes place near its boiling or condensation point. Instead, the water temperature follows a curve something like that shown in Fig. 3-4. At A, the air temperature increases; at B, the air temperature decreases. The water temperature "stalls" as it boils or condenses.

It takes a certain amount of energy to change a sample of liquid to its gaseous state, assuming that the matter can exist in either of these two states. In the case of pure water, it needs 540 cal (2260 J) to convert 1 g of liquid at +100°C to 1 g of vapor at +100° C. In the reverse scenario, if 1 g of pure water vapor at +100°C condenses completely and becomes liquid at +100° C, it gives up 540 cal (2260 J) of energy. This quantity varies for different substances. Scientists call it the *heat of vaporization* for the substance.

If we symbolize the heat of vaporization (in calories per gram or joules per gram) as h_v, the heat energy added or given up by a sample of matter (in calories or joules, respectively) as h, and the mass of the sample (in grams) as m, then

$$h_v = h/m$$

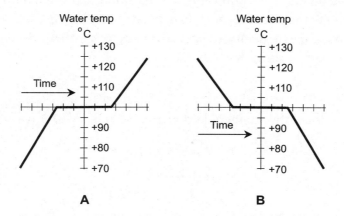

FIGURE 3-4 • Temperature-versus-time analyses of water as it boils and condenses. At A, liquid water takes in heat from the surroundings as it vaporizes. At B, water vapor gives up heat to the surroundings as it condenses. Note that the water temperature "stalls" as vaporization or condensation occurs.

Still Struggling

Certain substances known as *refrigerants* take in or release large amounts of heat when they change state. They're useful in systems that transfer heat energy from one place to another. We find the most common examples in electric "ice boxes" and small air-conditioning systems. Dozens of such chemicals exist. Scientists believe that some of them might contribute to environmental problems, such as *ozone depletion* and *climate change*.

PROBLEM 3-4

Suppose that a certain substance changes from liquid to vapor (or vice-versa) at +500°C under normal atmospheric pressure. Imagine a beaker containing 67.5 g of this substance entirely in liquid form at +500°C. If the heat of vaporization is 845 cal/g, how much heat energy, in calories, will this sample of matter absorb if it completely evaporates? Round off the answer to three significant figures, and express it in scientific (power-of-10) notation.

✔SOLUTION

We have quantities in grams for m and calories per gram for h_v. We must manipulate the above formula to express the heat h (in calories) in terms of the other given quantities. We can do this task with a little algebra, multiplying both sides by m to get

$$h = h_v m$$

When we input the numbers, letting $h_v = 845$ and $m = 67.5$, we can calculate

$$h = 845 \times 67.5$$
$$= 57{,}038 \text{ cal}$$
$$= 5.70 \times 10^4 \text{ cal}$$

PROBLEM 3-5

Imagine a substance that changes from vapor to liquid (or vice-versa) at +15°C at normal atmospheric pressure. Imagine an enclosed system that contains exactly 2 kg of this substance, entirely in vapor form, at +15°C. If the heat of vaporization is 8500 J/g, how much heat energy, in joules, will this sample of matter give up if it completely liquefies? Express the answer to two significant figures in scientific notation.

 SOLUTION

First, we must convert the mass of the sample, 2 kg, into grams. We know that 1 kg = 1000 g, so 2 kg = 2000 g. Then we can take advantage of the same formula that we used to solve Problem 3-4:

$$h = h_v m$$

We plug in the numbers, letting h_v = 8500 and m = 2000, and calculate to obtain

$$h = 8500 \times 2000$$
$$= 17,000,000 \text{ J}$$
$$= 1.7 \times 10^7 \text{ J}$$

PROBLEM 3 - 6

Convert the amount of heat energy derived in the solution to Problem 3-5 into British thermal units (Btu). Call this figure h_{Btu}. Round off the answer to the nearest 100 Btu.

 SOLUTION

From Chapter 1, we remember that 1 Btu = 1055 J. Therefore, we must divide the above result by 1055; then we can round off to the nearest 100 Btu, getting

$$h_{Btu} = 1.7 \times 10^7 / 1055$$
$$= 16,100 \text{ Btu}$$

Electric Heat Pumps

An *electric heat pump* uses electricity to operate a machine that transfers thermal energy from one place to another. The term "pump" comes from the fact that the device uses electricity to move heat energy from place to place instead of directly generating the heat energy.

Air-Exchange Heat Pump

Figure 3-5 shows the basic components of an *air-exchange heat pump*, also called an *air-source heat pump*, operating to transfer heat energy from the outdoor environment to the indoor environment (heating mode). The fan blows outdoor air through a coil that contains a refrigerant. As the refrigerant passes through the outdoor coil, depressurization and evaporation occur, causing the refrigerant to absorb heat energy. This process can occur even if the outdoor temperature is quite cool. The fluid then passes through pipes (shown as solid lines) to the indoor coil, where the refrigerant undergoes compression and condensation, causing it to release the heat energy it took in from the outside. The indoor coil becomes considerably warmer than ordinary room temperature; in fact, it can warm to as much as +35°C (+95°F) as it passes through the indoor coil. This warm air flows into the ductwork and circulates throughout the house.

Figure 3-6 portrays the same heat pump operating to transfer heat energy from the indoor environment to the outdoor environment (cooling mode). The

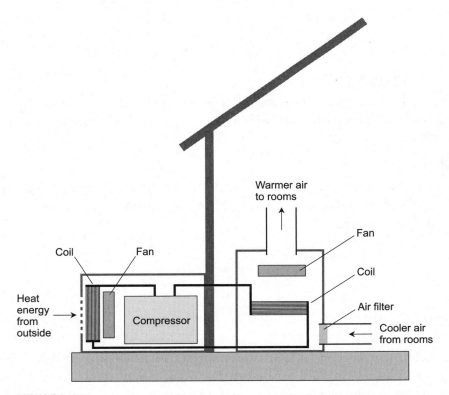

FIGURE 3-5 • An air-source heat pump, operating to transfer heat energy from the outdoor environment into a building.

fan blows indoor air through a coil that contains a refrigerant. As the refrigerant circulates through the indoor coil, depressurization and evaporation occur, so the refrigerant absorbs heat energy, thereby chilling the air that moves past the coil. The cooled air then flows into the ductwork and circulates throughout the house. The chilling process can also remove some excess humidity from the indoor air, causing the indoor coil to "sweat." The heated fluid passes through pipes (shown as solid lines) to the outdoor coil. In the outdoor coil, the refrigerant undergoes compression and condensation, causing it to give up the heat energy that it acquired from indoors. The outdoor fan blows warm air into the external environment.

Ground-Source Heat Pump

Some heat pumps extract thermal energy from beneath the earth's surface rather than from the outside air, and transfer this energy into a house or building. Figure 3-7 illustrates the principle. Basically, it's a modified air-exchange

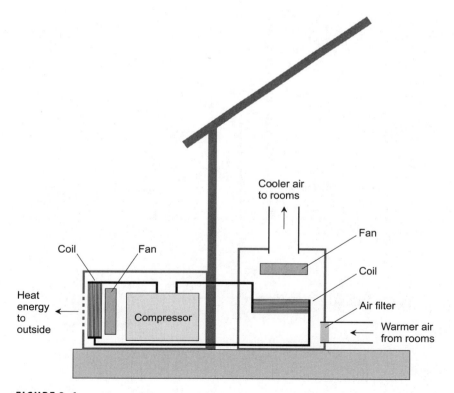

FIGURE 3-6 · An air-source heat pump, operating to transfer heat energy from inside a building to the outdoor environment.

system. The outdoor coil resides underground, so the system requires no outdoor fan. In some arrangements, the outdoor coil can be placed near the bottom of a deep pond or lake. Heat transfer occurs by conduction from the earth to the coils. The system shown in Fig. 3-7 constitutes a *ground-source heat pump*, also called a *geothermal heat pump*.

In some locations, the earth temperature is quite high, even at shallow depths. Saratoga, Wyoming, and Hot Springs, South Dakota, offer locations in the United States with plenty of available *geothermal heat* despite their severe

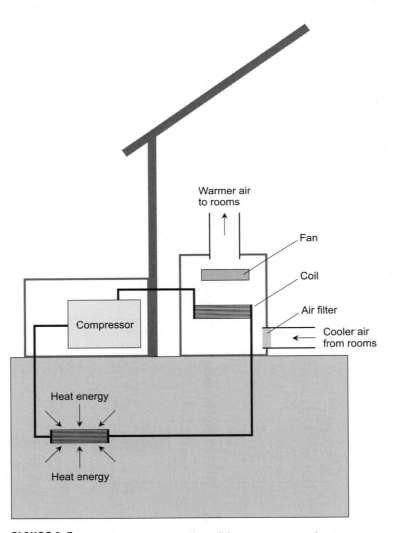

FIGURE 3-7 · A ground-source (geothermal) heat pump, operating to transfer heat energy from the earth into a building.

winters. Environmentalists often cite Iceland as another good example. In locations such as these, a ground-source heat pump can function at much lower outdoor air temperatures than an air-exchange heat pump can.

TIP   *Ground-source heat pumps can operate to cool the indoor environment during the summer in most locations. However, in some places (such as those mentioned above) the subsurface temperature is high. While this state of affairs is ideal for home heating with ground-source heat pumps, it doesn't lend itself to interior cooling with that technology.*

? Still Struggling (or Shivering?)

In locations where subsurface temperatures are hot enough even at shallow depths, a network of pipes can replace the outdoor coil, buried deep enough to allow heating of water that a mechanical pump (similar to the type of pump that brings water up from a well) can circulate through the house. That type of heating system is perhaps the most efficient possible; the only component that consumes any energy is the water pump, similar to the one that you'd find in an ordinary well.

Advantages of Heat Pumps

- Heat energy contained in the outdoor environment constitutes a renewable and practically unlimited resource.
- For heat pumps operating in cooling mode, the outdoor environment serves as an "infinite heat sink."
- Air-source heat pumps can heat the indoor air quite well, as long as the outdoor temperature remains above roughly +4°C (+39°F). A good heat pump can transfer more heat energy than it demands from the utility, so you'll sometimes hear a vendor claim that a heat pump's "efficiency" exceeds 100 percent. Actually, the vendor refers to the *coefficient of performance* (COP): the ratio of thermal energy transferred to the input energy that the system needs to do the job.
- Because heat pumps work best when called upon to provide a constant indoor temperature, you don't have to change the thermostat setting unless you plan to stay away from your house for several days.

- The exhaust from a heat pump is either cold or warm air, depending on the mode. The system produces no CO or other noxious gas. (However, some pollution does result indirectly at the distant electric power plant if it burns fossil fuels.)

- Ground-source heat pumps with sufficiently deep outdoor coil systems can function efficiently even in places where winters grow severe. At a depth of several meters beneath the surface, the temperature remains constant all year round—at least +10°C (+50°F) in most locations.

Limitations of Heat Pumps

- Air exchange heat pumps work well in the heating mode if the outdoor temperature exceeds about +4°C (+39°F). If the outdoor air gets colder than that, it doesn't contain enough thermal energy to allow efficient operation.

- In older systems that use *chlorofluorocarbon* (CFC) refrigerant compounds, the potential for *ozone depletion* exists, and that's a major issue! A small amount of CFC can destroy large numbers of ozone molecules. Ozone helps to shield the earth's surface from excessive solar ultraviolet radiation.

- In the heating mode, the air comes out of a typical heat pump near +35°C (+95°F). That's warmer than the typical indoor environment, but it won't heat up a cold house very fast.

- Heat pumps, especially deep ground-source systems, can cost a lot of money to install. It will take several years to recover the up-front expense in terms of the month-to-month savings that you realize, compared with the cost of running the system (however antiquated) that the heat pump replaced.

? Still Struggling

All humanmade heating or cooling systems act against the natural heat entropy that takes place as temperatures gradually equalize throughout the universe. In a simplistic sense, our furnaces and air conditioners impose small-scale order in a gigantic thermodynamic system that relentlessly strives for overall chaos, a process that some scientists call the *heat death of the cosmos*.

QUIZ

Refer to the text in this chapter if necessary. A good score is eight correct. You'll find the correct answers listed in the back of the book.

1. Expressed to the nearest degree Celsius, −10°F equals
   A. +14°C.
   B. −23°C.
   C. −76°C.
   D. −100°C.

2. Expressed to the nearest degree Fahrenheit, +25°C equals
   A. +112°F.
   B. +97°F.
   C. +46°F.
   D. +77°F.

3. Expressed to the nearest degree Fahrenheit, 233 K equals
   A. −40°F.
   B. −15°F.
   C. +40°F.
   D. +133°F.

4. Expressed to the nearest kelvin, standard temperature for scientific purposes equals
   A. 0 K.
   B. 100 K.
   C. 273 K.
   D. 373 K.

5. A 100-g sample of matter changes from all liquid to all vapor at a constant temperature of +278°F (which happens to be its boiling point) and standard atmospheric pressure. If the substance has 126 cal/g heat of vaporization, how much heat energy will this sample of matter absorb if it completely evaporates while remaining at +278°F?
   A. 1.26 cal
   B. 12.6 kcal
   C. 794 cal
   D. 7.94 kcal

6. Consider the same 100-g sample of matter, having all boiled off into vapor and still at a temperature of +278°F. Suppose that we allow the vapor to give up heat energy until it completely condenses, and then stop the process to prevent the temperature of the liquid from falling below +278°F. How much energy will the sample have lost?
   A. 1.26 cal
   B. 12.6 kcal
   C. 794 cal
   D. 7.94 kcal

7. **In the heating mode, and in a practical real-life situation, an air-exchange heat pump will work at its best when the outdoor temperature is**
   A. over 277 K.
   B. under 277 K.
   C. over 100°C.
   D. under 0°F.

8. **If you hear a physicist or engineer tell a lay audience that a certain heat pump operates at "250 percent efficiency," she's probably saying (in oversimplified terms) that**
   A. she doesn't know how heat pumps work, and she doesn't have a clue about basic physics principles, either.
   B. over any given period of time, the system requires more electrical energy to operate than the amount of heat energy that it transfers.
   C. the system can theoretically heat the indoor air to 2.5 times (250 percent of) the absolute temperature of the outdoor air.
   D. the system can transfer more heat energy into the house from the outdoor environment than it demands from the electric utility.

9. **When supplied with plenty of current, an electric resistance heating element gives off mostly**
   A. visible-light rays.
   B. ultraviolet rays.
   C. infrared rays.
   D. All of the above, in equal amounts

10. **Suppose that a certain heating element has a resistance of 11.7 ohms, and we provide it with AC electricity at 117 V RMS. How much power does the element dissipate?**
    A. 1170 Btu/h
    B. 3990 Btu/h
    C. 343 Btu/h
    D. 3410 Btu/h

chapter 4

Passive-Solar Heating

Solar energy can heat buildings in locations that receive reasonably abundant sunshine, even if their winters grow quite cold. In the United States, the West and South have sunnier weather in general than the North and East have, although localized exceptions exist. Passive-solar heating works better in Utah or Wyoming, for example, than it does in Michigan or Maine.

CHAPTER OBJECTIVES

In this chapter, you will

- Learn to harness the sun's energy and reduce your home heating costs.
- Discover how massive objects can absorb, retain, and release solar energy to keep you warm at night.
- See how forced-air methods can work with passive-solar systems to distribute heat energy throughout your house.
- Find out how water can store, release, and distribute the sun's heat energy.
- Evaluate the pros and cons of sidehill construction for energy conservation in passive-solar systems.

Sunnyside Glass

You can harness the sun's heat energy by placing large windows on the equator-facing side of a building, in conjunction with heat-absorbing and heat-retaining floors, walls, and furniture. You can also install windows in steeply pitched roofs, facing generally toward the equator (south in the Northern Hemisphere and north in the Southern Hemisphere), and with unobstructed views of the sky.

How It Works

When the sun shines through windows, visible light and shortwave IR rays penetrate the glass easily. Dark objects in the room absorb this energy and reradiate it in the form of longwave IR, as shown in Fig. 4-1. The window glass is less transparent to longwave IR than to visible light and shortwave IR. The longwave IR rays reflect from light-colored walls, ceilings, furniture, and the window panes, "bouncing around" and further heating up dark objects in the room. Longwave IR continues to radiate from dark objects in the room for awhile after the sun has gone down or the weather has turned overcast. Unless those dark longwave-IR-radiating objects are substantial, this effect doesn't last long. But if they're massive and numerous, the effect can continue for quite awhile.

Engineers call the scheme of Fig. 4-1 *basic passive-solar heating*. It will work in any location where the sun shines on most days, even if the outside temperature drops to frigid levels during part of the year. It will work in any room with large equator-facing windows. During the night, vertical blinds or heavy

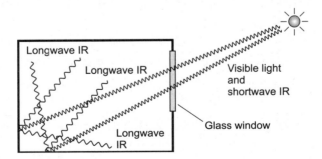

FIGURE 4-1 · When sunlight shines through a window, shortwave IR and visible light penetrate the glass, which is opaque to longwave IR that gets reradiated from objects in the room. That's why a "sunny room" warms up.

curtains can minimize heat loss through the windows, which can occur by conduction and convection as a result of contact between the air in the room and the cold glass.

Basic passive-solar heating can function even in climates that most people consider severe in the winter, as long as the sun shines enough, and as long as it grows intense enough (it must rise at least 10 or 12 angular degrees above the horizon at noontime). The Bighorn Basin of Wyoming offers a good example of such a place. Basic passive-solar heating would prove far less practical in a region such as the coast of Oregon, where, although winter temperatures don't get terribly cold, the sky remains overcast most of the time.

Most homes can take advantage of sunshine in windows that face toward the equator, and to some extent windows that face in any direction except directly toward the pole. Even a small window, with direct midwinter sun streaming in, can produce considerable warmth. The houses that work best for basic passive-solar heating, in the absence of modification, are structures that have a long side facing directly toward the equator, with large windows and no obstructions.

TIP *If you open the blinds and curtains when the sun shines on the windows, and close them when the windows don't receive direct sunlight, you can save quite a lot of money on your heating bill over the course of a year.*

Advantages of Basic Passive-Solar Heating

- It costs very little to install and almost nothing to maintain. You don't have to make any major structural changes to your house. A dark "couch throw" or two, some dark floor rugs, and some efficient blinds or curtains can make a big difference.

- A little personal behavior modification can result in significant energy savings, even in the absence of any other changes. Open the blinds or curtains when the sun shines into exposed windows, and close them the rest of the time.

- Snow on the ground can improve the effectiveness of passive-solar heating because it increases the amount of solar energy that enters the windows on sunny winter days (snow reflects sunlight). Snow, of course, tends to coincide with the heating season.

- Bright sunlight, even if it shines through glass, can elevate some people's moods if they suffer from *seasonal affective disorder* (the "winter blues").

Limitations of Basic Passive-Solar Heating

- Sunshine on furniture and carpeting can cause color fading over time. Moreover, the effect occurs unevenly because some portions of a room receive more direct sunlight than other parts.

- A room might overheat during the day if it receives abundant, direct sunshine for several continuous hours. If you want to stay warm at night, you'll feel the temptation to "get ahead of the game," driving the afternoon temperature in some rooms to uncomfortably high levels.

- Some buildings aren't "passive solar friendly." These include structures shaded by various objects, such as evergreen trees or other buildings, structures on hillsides facing toward the pole, and buildings in which few or no windows face toward the equator. In these cases, you can't expect passive-solar heating to work well.

- The same windows that allow sunshine in during the day, thereby heating the interior, can lose heat at night unless they're well designed. Thermal insulation such as caulking, along with multipane glass, can help. You must also remember to close the curtains or blinds at night.

PROBLEM 4-1

When I was a child, my mother would close the curtains over the big picture window in the living room during the daylight hours, especially in winter when low-angle sunshine would otherwise fill the room. She feared that the sunlight would fade the furniture and carpet. How can I minimize this fading and still take advantage of passive-solar heating?

SOLUTION

You can reduce color fading by installing double-pane or triple-pane glass windows. Most fading results from UV radiation, not IR or visible light, and glass doesn't transmit UV very well. However, you can't expect to avoid fading altogether. "Couch throws" and inexpensive rugs can help to protect your furniture and carpet, but some people think those protective devices look "cheap." (Of course, you can reject passive-solar technology and save on your heating costs in some other way, for example by installing a heat pump. Then you can close your curtains or window blinds all day long and prevent color fading entirely.)

Thermal-Mass Heating

If used without any other heating method, passive-solar heating allows fluctuations in temperature between sunny periods and cloudy periods, or between day and night. To smooth out these variations, you can add *thermal mass* to an existing structure or build it into a new structure. Thermal mass produces *thermal inertia*, so the indoor air stays at a fairly constant temperature despite large short-term variations in available solar energy.

Floor and Wall Slabs

Substances with high *material density* (a lot of weight for the volume) make the best thermal masses. Concrete is one of the most effective materials for this purpose. Brick, adobe, and rock (such as limestone) also work well for building thermal masses.

Figure 4-2 shows thermal mass in the floor and interior walls of a room that receives direct sunlight through a glass window. In the ideal case, the sun shines

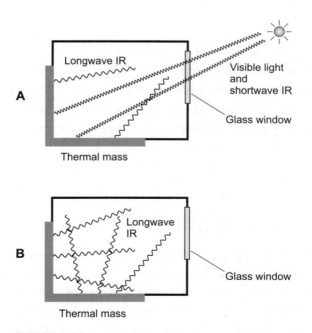

FIGURE 4-2 · Thermal mass in interior walls and floor. When the sun shines (A), the mass absorbs heat energy as visible light and shortwave IR. During cloudy periods or at night (B), the mass radiates the heat energy as longwave IR.

directly on the surface of the thermal mass to the greatest extent possible. The surface of the mass has a dark color, so it absorbs most of the visible-light and shortwave IR energy that strikes it. Bare concrete doesn't appeal to very many people as an interior wall or floor finish, although some people find it attractive if it has special finishing or painting. Interior brick, with concrete behind it, can make a presentable room interior.

When the sun shines into the room (Fig. 4-2A), the thermal mass absorbs energy primarily from visible light and shortwave IR. The thermal mass radiates some longwave IR, which gradually dissipates in the room, warming the walls, floor, ceiling, and furniture, and ultimately, the air by conduction. When no solar input exists (Fig. 4-2B), the thermal mass slowly gives up its heat energy in the form of longwave IR. The thermal mass also conducts heat to the air by direct contact.

A thermal mass installed in a floor should be insulated from the earth below to prevent the loss of thermal energy to the ground by conduction. Thermal masses don't work well in outside walls for the purpose of heating a building with passive-solar energy, unless those walls have exposed interior surfaces and substantial insulation exists between the thermal mass and the exterior surface.

TIP *In the extreme, you can use reinforced and poured concrete to build your entire house! This method of construction finds favor in locations prone to hurricanes in which flying debris can smash an ordinary house to pieces. As an incidental asset, a house of this sort keeps out external noise quite well. It's expensive, but some people consider the benefits worth the up-front cost.*

? Still Struggling

Thermal inertia increases in direct proportion to the total thermal mass. You can expect to get the best results with large thermal masses if long periods pass with little or no sunshine. The more thermal mass that you install, the longer it takes to heat up, but it will release heat for a longer time, too. In some passive-solar homes, concrete slabs as thick as one meter (1 m) or more serve as floors and subterranean walls to maximize the thermal mass.

Roof Windows and Ceiling Overlays

You can obtain passive-solar heating by installing windows in a steeply pitched roof, along with a thermal mass on the attic floor. The bottom of the thermal mass forms the ceiling of the living space. Direct sunlight warms the thermal mass (Fig. 4-3A) by irradiating it from the top with visible light and IR. The thermal mass radiates longwave IR downward into the living space by day (Fig. 4-3A) as well as during the night or on cloudy days (Fig. 4-3B).

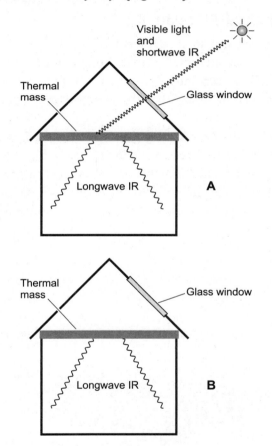

FIGURE 4-3 · Thermal mass in ceilings (attic floors), along with windows on the sunny side of a steeply pitched roof. When the sun shines (A), the mass absorbs energy as visible light and shortwave IR, and also radiates longwave IR into the rooms below. During cloudy periods or at night (B), the mass continues to radiate longwave IR into the living space.

If you use basic passive-solar heating in your living space in addition to thermal mass in the attic floor, you can reap the benefits of both methods. An even better scheme involves the use of additional thermal masses in the floor and/or walls of the living space. However, the passive-solar heating method shown in Figs. 4-3A and 4-3B can function all by itself, even if the living space has few windows. The main challenge is constructing the building so the exterior walls can support the massive attic floor.

Passive-Solar Forced-Air Heating

Envision a thermal mass in an attic floor, made of concrete blocks laid so that their openings are horizontal and so that the openings line up for the entire width of the attic. This arrangement creates airways through the thermal mass (Fig. 4-4). Blowers can drive air from the living space, through these airways, and back into the living space.

In a *passive-solar, forced-air heating* system of this type, the air intake vent, along with the set of blowers, resides in one of the walls. The air outflow vent resides in the opposite wall. A second set of blowers in the airway near the outflow vent can enhance the circulation. A network of ducts ensures that the air passes uniformly through all the airways produced by the openings in the blocks.

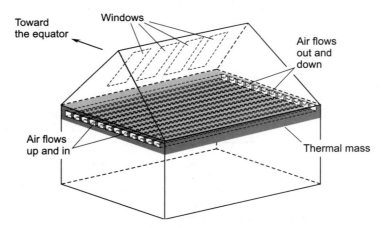

FIGURE 4-4 · Air passageways in a thermal mass facilitate forced-air heating with a passive-solar system. Outlines represent the perimeters of the interior surfaces. (This diagram doesn't show the blowers and ductwork.)

TIP *In a well-designed, passive-solar, forced-air heating system, thermal energy moves from the blocks to the air as the air flows through the ducts. In addition, heat energy propagates into the living space by means of longwave IR radiation downward from the thermal mass in the ceiling.*

Advantages of Thermal-Mass Heating

- Thermal mass reduces fluctuations in temperature between day and night, or between sunny days and cloudy days.

- If a prolonged cloudy spell occurs, the heat acquired during sunny periods will last longer in a building with more thermal mass than in a building with less.

- The inclusion of thermal mass in new construction can result in a building more likely to withstand severe weather, particularly high winds accompanied by flying debris.

- A building constructed with substantial concrete, stone, or brick offers better resistance to fire than a conventional frame building does.

- Thermal mass can provide some acoustic insulation (soundproofing) between the rooms or levels of a house, and between the interior and exterior.

Limitations of Thermal-Mass Heating

- The cost of constructing a new home with significant thermal mass can prove prohibitive to many people. Shortages of concrete sometimes occur, and whenever that happens, the cost of concrete rises.

- Retrofitting a conventional building with thermal mass can be complicated and expensive. If not done properly, such retrofitting can also create a danger for the occupants.

- Some people consider thermal mass esthetically unattractive, no matter how well-disguised or adorned.

- A building with a lot of thermal mass "feels heavy" when you're inside it! Some people like this feeling; others find it oppressive.

PROBLEM 4-2

Why doesn't wood work well as a thermal-mass medium? It absorbs and radiates heat, doesn't it? Why should density make any difference in the effectiveness of a particular material for use as a thermal mass?

SOLUTION

As the density (in kilograms per cubic meter) increases, so does the number of subatomic particles (particularly protons and neutrons) per unit volume, in general. The more "stuff" that exists in a certain volume of space, the more heat energy it can retain, and the longer it takes to lose it all. This fact explains why, for example, you'll often find stones in saunas and sweat lodges. Wood has relatively low density (not much "stuff" per unit volume), so it doesn't work very well as a thermal-mass medium.

Solar Water Heating

If you've ever lived near a lake, you know that the water temperature changes very little from one day to the next, even if weather conditions vary drastically. This state of affairs prevails because water has excellent thermal mass characteristics. Have you noticed that a swimming pool takes a long time to get warm from the sun, and an equally long time to cool down on cloudy days or in cold weather? Water "holds heat" quite well.

How It Works

You can heat water by passing it through pipes embedded in black metal panels under glass, exposed to direct sunlight. Engineers call these sealed panel-and-glass assemblies *flat-plate collectors* or *flat-panel collectors*. Ideally, you'll mount them on a steeply pitched, equator-facing roof, secured to keep wind from breaking them loose. Alternatively, you can mount them with their flat surfaces perpendicular to a line running up into the sky toward the noonday, midwinter sun.

You can install flat-plate collectors in a yard or field, or on a flat or moderately pitched roof. The combination of the *greenhouse effect*, in which longwave IR radiation gets trapped beneath the glass, and the fact that the black panels absorb (and convert into heat) all the radiant energy that strikes them, can produce panel temperatures higher than the boiling point of water on sunny days.

If you want to use solar-heated water for washing or bathing, an indirect heating scheme works better than passing the water through flat-plate collectors directly. Special *heat-transfer fluid* allows for the storage of heat energy in a hot-water reservoir resembling the tank of a conventional water heater. The heat-transfer fluid won't freeze in the pipes or panels during frigid overcast or nighttime conditions as water will do because heat-transfer fluid acts as anti-freeze. As people in frigid-winter country know, water expands when it freezes solid, damaging pipes and water fixtures that remain filled with it.

In a solar water-heating system, the water-storage tank should have a capacity of at least several hundred liters (about 100 gallons or more) so that it will have substantial thermal mass when full. The tank should have a covering of thermal insulation to keep it from giving up thermal energy to the surrounding air. Figure 4-5 shows the basics of a solar water-heating system that uses flat-plate collectors.

TIP *When considered as a whole, the arrangement shown in Fig. 4-5 constitutes an active system because of the presence of the pump for the heat-transfer fluid, which operates on electricity.*

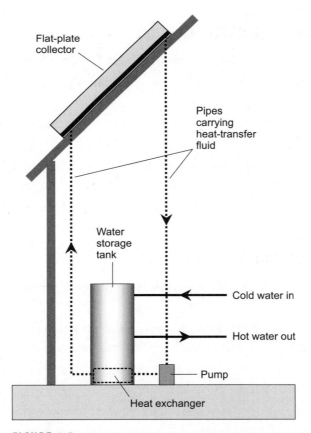

FIGURE 4-5 · Functional diagram of a water-heating system that uses flat-plate collectors and heat transfer fluid.

? Still Struggling

If you want to use the energy derived from flat-plate collectors to heat your home's interior, you won't need a water-storage tank. The heat-transfer fluid can pass directly through the collectors, pumped through pipes embedded in concrete thermal masses in the floors or walls. In this case, the flat-plate collectors take the place of a gas, propane, or oil-fired boiler, and the heat-transfer fluid takes the place of the water. The circulation pump shuts down during the night or on cloudy days, but the thermal mass continues to radiate thermal energy into the living space.

Advantages of Solar Water Heating

- Only the pump requires electricity, and it doesn't consume much power.
- It's a closed system, neither leaking nor requiring the introduction of outside substances.
- It generates no pollutants, except indirectly as a result of the electricity used by the pump (if the power plant burns fossil fuels).
- It makes almost no noise. You can locate the small, relatively quiet pump motor in an out-of-the-way place, such as a basement or broom closet.
- The system has few moving mechanical parts to break down.

Limitations of Solar Water Heating

- This type of system will not work for prolonged periods of cold, cloudy weather.
- Flat-plate collectors can suffer damage or destruction from hail storms, ice buildup, falling tree limbs, or other adverse events.
- If you don't mount the flat-plate collectors flush with a roof or other surface, a high wind can rip them loose.
- You must keep the flat-plate collectors clear of snow. If you've mounted them on a rooftop, you'll find this task inconvenient.

- Systems for heating a living space require many flat-plate collectors, driving up the expense, especially in regions where the weather grows cold often.

- Flat-plate collectors don't work well in locations where the sun rarely shines.

 PROBLEM 4-3 _____

Can a system, such as the one described in this section, serve to heat a swimming pool?

✔ SOLUTION _____

Flat-plate collectors have been used for decades to heat swimming pools. If the pool is located outdoors, the pool water directly circulates through the collectors. In extremely hot locations, you can reverse the cycle to cool the pool water, using the collectors to radiate thermal energy into the atmosphere during the night (leaving the system off in the daytime).

Sidehill Construction

An ideally designed passive-solar building lies "square with the compass," so the long sides of the structure face north and south. The equator-facing side has the most window area, and the pole-facing side has the least window area (or none at all). The pole-facing side has excellent thermal insulation, so it loses minimal energy to the outdoor environment. All these criteria together have given rise to the concept of *sidehill construction*, in which a structure is built into the earth on the equator-facing face of a natural or humanmade hill.

TIP *During the 1800s and early 1900s, pioneers in the American Midwest and Great Plains took advantage of sidehill construction methods, building their homes partway into the south-facing sides of whatever small hills they could find.*

How It Works

Figure 4-6 shows the basics of a passive-solar home built with sidehill design. Windows in the equator-facing side of the pitched, asymmetrical roof produce

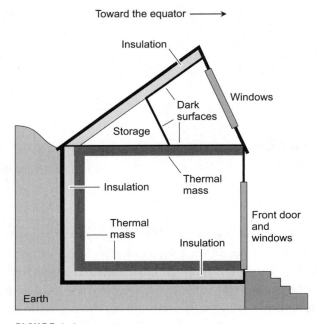

FIGURE 4-6 · Sidehill building design with thermal masses and passive-solar windows.

heat by greenhouse effect during the daylight hours. At night, automatically controlled blinds close the roof windows, minimizing heat loss.

Thermal masses in the attic floor (the ceiling of the main living space), as well as in the pole-facing wall and the floor, facilitate heat retention so that the home remains warm at night and on cloudy days. Insulation behind and beneath the wall and floor thermal masses minimizes conductive heat loss into the earth.

Figure 4-6 does not show the east and west walls of the house; both of those walls lie about half below the surface and half above. These walls can contain thermal mass along with exterior insulation, similar to the design of the pole-facing wall.

TIP *East- and west-facing windows can admit some sunshine into the living space, but not as much as equator-facing windows do (obviously). The east-west windows, like all the windows in the passive-solar, sidehill house, should have curtains or blinds that you can close when sunlight doesn't strike them.*

Advantages of Sidehill Construction

- With good thermal insulation, sidehill construction offers superior heat retention.

- Sidehill construction offers protection from cold polar winds, and to some extent, from strong easterly or westerly winds.

- Sidehill houses let in less external noise than conventional frame houses do.

- Sidehill houses have less exterior surface area, which translates to less exterior maintenance work, compared with other types of houses.

- You can expect a sidehill home to stay cooler in the summer than a conventional home does. You can open the windows at night. During daylight hours, you can close the windows and draw the blinds on the side(s) of the house exposed to sunshine.

Limitations of Sidehill Construction

- Earth movement can pose a problem in some locations. Over time, an unstable hill will destroy a home built into the slope.

- To prevent flooding of the interior during spring snowmelt or heavy rain storms, you must ensure that the home has adequate drainage.

- Walled-off rooms in the pole-facing side of the building will receive little or no natural light unless you place translucent windows in the interior walls, but these windows can allow thermal energy to escape.

- The roof of a sidehill building can present a liability issue if someone ventures onto the roof from the pole-facing side and then falls off one of the other sides.

- The accumulation of *radon gas* can pose a problem because a large part of the building lies below ground. Long-term exposure to radon gas can increase the risk of certain health problems, notably lung cancer. You must ensure that radon levels can't rise to unhealthful levels in the living space, a task that can prove quite expensive to carry out.

PROBLEM 4 - 4

My prospective home site in Wyoming offers a fabulous view toward the north. Even though the site lies on the south (equator-facing) side of a hill, I can still

see the mountain peaks to the north. How can I build an energy-efficient home in this situation and at the same time get a view toward the north?

SOLUTION

You can install small windows in the north side of the house. The earth does not have to come all the way up to the level of the roof on that side. You'll have to locate the windows near the ceiling, forcing you to look up to see through them; but that should work out okay because the line of sight runs upslope anyway. If you have a professional installer add tinted plastic to the surfaces of these windows, use multiple-pane glass, and employ efficient blinds or curtains to minimize radiation loss at night, you can enjoy that northerly view during daylight hours without sacrificing too much efficiency.

QUIZ

Refer to the text in this chapter if necessary. A good score is eight correct. You'll find the correct answers listed in the back of the book.

1. **If you want to minimize color fading of your carpeting or furniture in a passive-solar house, you can**
   A. install double-pane or triple-pane windows.
   B. use cheap rugs and "couch throws" to cover surfaces prone to fading.
   C. forget about passive-solar heating and use another scheme to save energy.
   D. Any of the above

2. **With a passive-solar heating system,**
   A. some rooms can get too hot in the daytime.
   B. you don't have to spend much money to get results.
   C. windows can lose thermal energy at night.
   D. All of the above

3. **In a solar water heating system, the use of heat-transfer fluid in the flat-plate collectors can**
   A. prevent system freeze-up in the winter.
   B. optimize the transfer of thermal energy to and from the earth.
   C. facilitate direct heating even during hours of darkness.
   D. All of the above

4. **In a passive-solar home located in South Dakota, USA, the side with the *most* window area should face**
   A. north.
   B. south.
   C. east or west.
   D. in any direction; it doesn't matter.

5. **In a passive-solar home located in South Dakota, the side with the *least* window area should face**
   A. north.
   B. south.
   C. east or west.
   D. in any direction; it doesn't matter.

6. **Passive-solar heating can work where winters grow frigid, provided that**
   A. the days remain two or more hours longer than the nights all year round.
   B. snow doesn't fall too often, and when it does, it never gets very deep.
   C. the sun shines a lot, and it attains reasonable intensity at midday.
   D. All of the above

7. **Which of the following situations or phenomena can cause trouble with a sidehill-constructed home in some locations?**
   A. Lack of precipitation
   B. Excessive sun exposure
   C. Radon gas
   D. Extremely hot summers

8. **Which of the following materials should we expect to work best as a thermal mass?**
   A. Wood
   B. Plastic
   C. Air
   D. Concrete

9. **Why should a thermal-mass floor be insulated from the earth below in a system designed for heating?**
   A. To keep frost in the ground from thawing and causing structural damage to the house in the winter.
   B. To prevent conductive loss of thermal energy from the house's interior to the ground.
   C. To keep excessive subterranean heat from warming the house's interior at night or in the summer.
   D. The premise of this question is wrong! A thermal-mass floor should never be insulated from the earth below.

10. **In a passive-solar house with large windows, color fading of interior carpeting and furniture results primarily from radiant energy in**
    A. the visible spectrum.
    B. the ultraviolet (UV) spectrum.
    C. the infrared (IR) spectrum.
    D. all parts of the spectrum; it doesn't matter.

chapter 5

Alternative Indoor Climate Control

This chapter describes some not-so-common ways of heating and cooling a building. These schemes rarely work as the primary modes of indoor environment control, but they can off-load some of the burden from conventional systems.

CHAPTER OBJECTIVES

In this chapter, you will

- Learn how a wind turbine can provide direct "part-time" power to heating elements.
- See how a water turbine can directly power heaters, air conditioners, and appliances.
- Discover how solar cells can provide direct "part-time" power for indoor climate modification.
- Find out how thermal masses can help keep a house cool on hot days.
- Learn how evaporative systems can reduce cooling costs in desert regions.
- Compare the assets and limitations of subterranean living.

Direct Wind-Powered Climate Control

You can connect a *wind turbine* to an electric generator, which in turn can provide current to electric heating elements. With proper voltage regulation, this arrangement can provide supplemental heat for a modest home when the wind blows.

How It Works

Figure 5-1 illustrates a wind turbine, equipped with an electric generator, connected into a zone electric baseboard heating system. A voltage-regulation

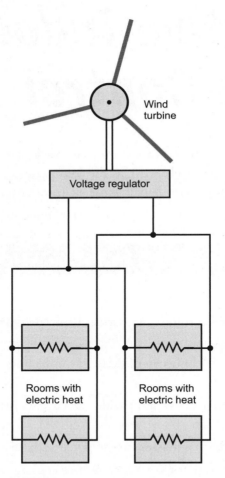

FIGURE 5-1 · A supplemental interior heating system that uses a wind turbine and voltage regulator connected into a conventional electric home heating circuit.

circuit maintains the system at or near 117 V AC, so the heating elements can operate as they normally would with the electric utility.

A medium-sized wind turbine designed for residential use can produce 12 kW of power on a day with moderate wind. That's the equivalent of eight electric space heaters, each rated at 1500 W. From Chapter 1, recall that 1 kW = 3410 Btu/h. Therefore, the wind turbine system of Fig. 5-1 can provide approximately $3410 \times 12 = 40{,}920$ Btu/h. (Let's round this off to 41,000 Btu/h). A gas furnace for a typical residential home produces 80,000 to 100,000 Btu/h when running "full blast." In theory, then, on a bitter-cold winter day with some wind, the system shown in Fig. 5-1 can supply about half of the energy necessary to keep a house warm.

A system like that shown in Fig. 5-1 depends on wind for its operation. Batteries of reasonable cost can't store the large amounts of energy required for home heating. In a location with little wind, this scheme won't prove cost effective. However, in some places, winters remain cold and windy for weeks or months at a time. Such places make good "proving grounds" for a system such as the one diagrammed in Fig. 5-1.

TIP *On a hot, windy day, a wind-powered system can operate an air conditioner or evaporative cooler. But in hot, calm weather, you'll need some other source of energy to cool your home.*

Still Struggling

The wind can serve as an excellent source of supplemental or intermittent electrical power, but wind doesn't lend itself well to use as a continuous, high-volume energy source. When properly exploited, wind energy can reduce dependence on fossil fuels. For the idealist, that's good news. But you probably shouldn't disconnect or uninstall an existing gas, propane, or oil furnace in favor of a system like that shown in Fig. 5-1. Later in this book, we'll take a closer look at the principles of wind power.

Advantages of Direct Wind-Powered Climate Control

- It can significantly reduce (but not eliminate) your reliance on conventional methods of home heating.

- It will function in some situations when you need heat, even if your normal home heating system won't work because of a loss of electric utility power.
- The system generates no greenhouse gases, particulates, carbon monoxide, ground contamination, or waste products.
- You can use the electricity produced by the wind turbine for general intermittent purposes, such as charging batteries. When you don't need heat, the electricity can provide interior lighting, and can operate other devices not sensitive to sudden voltage drops.
- If you want, you can use the wind turbine as the starting point for a stand-alone or interactive wind energy system.

Limitations of Direct Wind-Powered Climate Control

- When the wind doesn't blow at least a few miles per hour, a direct wind-powered system won't function.
- Wind turbines won't work properly if the wind is too strong. A well-designed turbine will "fold its propellers" and ride out the storm.
- A powerful wind storm, tornado, hurricane, ice storm, or lightning strike can destroy a wind turbine.
- The type of system described here will take a long time to pay for itself (and might never do so). Midsize residential wind turbines, including the cost of labor and ancillary materials such as a support tower, cost a lot of money.
- Some people dislike the way wind turbines look. Most wind turbines make noise when they run. For these reasons, you'll rarely see residential wind turbines in cities or suburbs.

PROBLEM 5-1

Will a traditional windmill operate as a wind turbine? When I drive around in the country, I occasionally see these old relics. Some of them appear to be in good physical condition.

SOLUTION

In any application where intermittent voltage reductions can be tolerated, and where you don't need a continuous source of electrical power, you can connect an old-fashioned windmill to an electric generator and voltage

regulator for supplemental power. However, old-fashioned windmills don't convert wind energy to electrical energy as efficiently as modern wind turbines do.

Direct Hydroelectric Climate Control

You can connect a *water turbine* to an electric generator, which can drive electric heating and cooling systems in much the same way as a wind turbine can. With sufficient water flow and proper voltage regulation, an arrangement of this kind can provide some of the power for climate control in a typical household.

How It Works

Figure 5-2 is a block diagram of a small water-driven energy system adapted for electric baseboard home heating. This assembly resembles the direct wind-powered system shown in Fig. 5-1, except that we replace the wind turbine with a water turbine. As with the wind system, a regulator circuit keeps the voltage near 117 V AC. A substantial system of this sort can provide enough power to run an electric air conditioner as well as major home appliances such as refrigerators.

A good water turbine, installed in a fast-moving stream or small river with sufficient vertical drop, can produce 20 kW of power on a reliable basis. Again, recall that 1 kW = 3410 Btu/h. Therefore, a substantial water turbine system can provide approximately $3410 \times 20 = 68,200$ Btu/h. (Let's round this off to 70,000 Btu/h). That amount of power can keep a small or medium-sized home comfortable in almost all types of weather, as long as the stream or river doesn't dry up or freeze solid.

Advantages of Direct Hydroelectric Climate Control

- Direct hydroelectric power can significantly reduce, and possibly eliminate, reliance on conventional methods of home heating and cooling.
- Water flow, unlike the wind, remains continuous (if the stream or river is large and fast enough).
- A direct hydroelectric system will function when the normal indoor environmental control system won't work because of a loss of electric utility power.

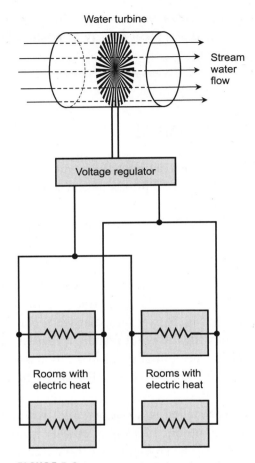

FIGURE 5-2 · An interior heating system that uses a water turbine and voltage regulator connected into a conventional electric zone heating circuit.

- The electricity produced by a water turbine can serve other purposes besides indoor climate control (lighting and power for small appliances, for example).
- You can modify a water turbine to power a stand-alone or interactive hydroelectric system of the sort described later in this book.
- Water turbines generate no greenhouse gases, particulates, carbon monoxide, ground contamination, or other waste products. A small amount of "thermal pollution" goes into the stream as a result of friction in the

bearings among the moving parts of the turbine, and also between the turbine and the stream water; but a residential-scale system doesn't produce enough of this "pollution" to pose a significant environmental problem.

Limitations of Direct Hydroelectric Climate Control

- Very few people live on properties with rivers or streams with enough flow to provide significant hydroelectric power.

- A small stream might periodically dry up in a prolonged drought, or freeze solid (from surface to bottom) in an extreme cold snap. Then, obviously, a water turbine won't work.

- A water turbine requires considerable water mass, along with a significant vertical drop, in order to provide enough power to heat a home. You might have to install a small dam on your property to make your system work to your satisfaction. That action could give rise to problems with environmentalists and regulatory agencies.

- A system, such as the one described here, will take a long time to pay for itself, and in fact, may never recoup all of the investment. A home hydroelectric system costs about the same as a wind-powered system that provides comparable output.

PROBLEM 5-2

A stream runs through my property. It's 10 m (33 ft) wide and 2 m (6.6 ft) deep in the middle. It flows well except in winter, when it freezes on the surface, although the water keeps flowing under the ice. The vertical drop, from the point where the stream enters the property to the point where it exits, only amounts to 0.5 m (approximately 18 in). Will this stream provide enough hydroelectric power to heat my home?

SOLUTION

You can get an engineer to evaluate the situation, but based on this information, your stream probably won't provide enough power to heat your home. Even so, a water turbine in this stream could produce a few watts of electricity on a reliable basis, enough to operate a few small lamps and to keep a notebook computer charged up.

Direct Photovoltaic Climate Control

Arrays of *photovoltaic* (PV) *cells* normally operate in conjunction with *storage batteries*, or else as a supplement to the electric utility. Nevertheless, you can make such arrays, called *solar panels*, drive a stand-alone system without batteries if you accept the fact that you'll only get power from them when the sun shines! You'll learn more about PV cells and photovoltaic systems in Chapter 12.

How It Works

Figure 5-3 is a simplified block diagram of a direct PV system for indoor environment modification. In bright sunshine, a single *silicon PV cell* produces

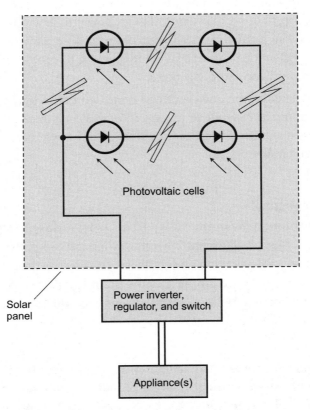

Photovoltaic cells

Solar
panel

Power inverter,
regulator, and switch

Appliance(s)

FIGURE 5-3 · A supplemental indoor environment control system that uses a solar panel, power inverter, voltage regulator, and switch connected to a small appliance such as a fan or humidifier.

approximately 0.5 volts of direct-current electricity (V DC). You can connect numerous silicon PV cells in a *series-parallel array* that provides 12 V DC or 24 V DC output at fairly high current in direct sunlight. Engineers call an array of this type a *solar panel*.

TIP *When you connect electrical components in series, you connect the right-hand end of one to the left-hand end of the next, arranging them like the links in a chain. When you connect components in parallel, you connect all the left-hand ends together and all the right-hand ends together, arranging them like the rungs of a ladder. When you connect components in series-parallel, you connect two or more series-connected sets in parallel, or else two or more parallel-connected sets in series, getting a matrix or array.*

Figure 5-3 shows only four PV cells (for simplicity), but it illustrates the basic series-parallel principle. A real system can contain hundreds of individual PV cells. A *solar module* of 53 silicon PV cells connected in series, each rated at 0.5 V DC, theoretically yields 26.5 V DC with the same maximum current output as a single cell. When you call upon the system to produce power, this figure drops to around 24 V DC. By connecting multiple 53-cell series modules in parallel to form a *solar panel*, you can obtain high current levels at the same voltage (in this case 24 V DC).

The output of the solar panel goes to a circuit called a *power inverter* that changes the low-voltage DC output of the solar panel into 117 V AC that can operate ordinary home appliances. The system includes a *voltage regulator* to ensure that the voltage remains fairly constant under conditions of varying solar intensity.

A well-designed direct PV system has an *automatic shutdown switch* that disconnects the solar panel if the daylight becomes too dim to properly operate appliances connected to it. If the system runs near peak capacity and the delivered current suddenly drops (a storm cloud moves in, for example), the switch will power-down the system until sufficient daylight returns.

TIP *Actual solar panels vary greatly in their designs and output specifications. Some of them provide voltages higher or lower than 24 V DC. Regardless of the voltage level, the maximum available power that you can get in bright, direct sunlight from a solar panel comprising multiple identical PV cells varies in direct proportion to the total surface area of the array.*

TIP   *Electric space heaters and air conditioners draw too much current to work with a solar panel of reasonable size. Theoretically, you can power-up such appliances with solar panels, but the panels would have to be so large that the benefit would probably not justify the cost.*

? Still Struggling

Electric fans, humidifiers, and evaporative coolers don't draw much current, so you can use them intermittently with a direct PV source of power. In summer in a place such as southern Arizona, for example, the hottest part of the day usually has bright sunshine. In a place like that, the system shown in Fig. 5-3 could operate a set of ceiling fans in a home or business. In a cold but sunny desert region such as northern Nevada or central Wyoming, the system could operate a humidifier to mitigate extreme low-humidity indoor conditions that prevail in winter.

Advantages of Direct PV Climate Control

- A direct PV system has no batteries and no complex tie-in to the electric utility. This simplicity minimizes the cost.
- Because of the system's simple nature, very little can go wrong with it, as long as you install it correctly and the solar panels don't get damaged.
- The system needs almost no maintenance, except perhaps for removing snow from the solar panels in the winter.
- You can upgrade a direct PV system to a stand-alone PV system (using storage batteries) or an interactive PV system (tied into the electric utility).

Limitations of Direct PV Climate Control

- A direct PV system will only operate on bright days (direct sunshine or light overcast).
- On a hot day without enough light to operate the system, or during a hot night, the system won't work.
- A sudden hail or wind storm can destroy a set of solar panels.

- If the solar panels get covered with snow, you must remove the snow manually in order for the system to function.
- The total current demanded by the appliances must never exceed the system's rated maximum deliverable current, even for a moment.

PROBLEM 5-3

I have plenty of real estate, and I can place a solar panel array of practically un-limited size on it. I don't care how much it costs. If I install a gigantic PV array, can I run an air conditioner or electric space heater using the direct PV scheme shown in Fig. 5-3?

✔ SOLUTION

In theory, you can do that. However, if you have that much money to spend, you might consider a hybrid system that makes use of wind power in addition to solar power. Later in this book, we'll take a look at small-scale hybrid systems.

Thermal-Mass Cooling

The same thermal inertia that allows dense solids, such as concrete, brick, and stone to keep a home warm can also keep an interior living space cool. Thermal masses can "store cold" as well as "store heat"! People take advantage of this capability in regions where summer heat presents a greater challenge than winter cold does.

How It Works

Figure 5-4 shows a cross-sectional view of a ground-floor room in a home that uses thermal mass for cooling. The mass lies under the floor and on one or more of the exterior walls. The home lacks insulation between the floor mass and the earth beneath, allowing heat energy to dissipate into the ground by conduction. The exterior walls are painted white on the outside, or covered with white siding, to minimize the heat that they absorb from direct sunlight.

If the window in the room (Fig. 5-4) faces in a direction where the sun regularly shines, you can close the window during the daytime (A) and open it up at

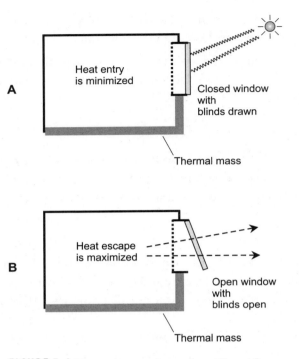

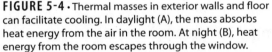

FIGURE 5-4 · Thermal masses in exterior walls and floor can facilitate cooling. In daylight (A), the mass absorbs heat energy from the air in the room. At night (B), heat energy from the room escapes through the window.

night (B). This maneuver keeps the hot outdoor air from entering during daylight hours, but allows warm indoor air to escape at night. You can close a set of curtains or blinds whenever the sun would otherwise shine into the room. When the window does not receive direct sunlight, you can open the curtains or blinds.

If the room's window happens to face toward the pole, then the curtains or blinds play a less important role, although in summer you might want to close them in the early morning and late afternoon hours when the sun occupies the poleward half of the sky.

TIP   *The ideal geometry for an arrangement of the sort shown in Fig. 5-4 comprises a sidehill home that opens up in the direction from which the sun shines the least, not the most! In the northern hemisphere, therefore, you'd do best to construct your sidehill home on a north-facing slope, and in the southern hemisphere, you'd want to build the structure on a south-facing slope.*

? **Still Struggling**

The thermal mass on an exterior wall heats up slowly during the day and cools down slowly at night. For that reason, the mass temperature doesn't depart much from the average air temperature over a 24-hour cycle. In the desert, this average might be quite warm, even though it is considerably below the peak afternoon temperature. A person accustomed to "meat-freezer-like" air conditioning might find this temperature a little uncomfortable in the height of the summer, but it's a huge improvement over the conditions that would exist in a conventional frame house without air conditioning. Arguably, it's also healthier than overdriven air conditioning that generates extreme contrast between outdoor and indoor temperatures.

Advantages of Thermal-Mass Cooling

- The proper installation of thermal mass, and the proper use of windows and curtains or blinds, can keep a home from growing dangerously hot, no matter how high the outdoor temperature gets.

- The use of thermal mass for cooling costs far less on a day-to-day basis than the use of active air conditioning. In a hot, dry place, you might fully recoup your initial cash outlay within a few years.

- If a prolonged hot spell occurs, heat energy will build up more slowly in a house with a lot of thermal mass than in a house with relatively little thermal mass.

- The inclusion of thermal mass in new construction can produce a structure more likely to withstand high winds than a "stick-built" house.

- A house constructed with substantial concrete, stone, or brick offers better resistance to fire than a wood frame house does.

Limitations of Thermal-Mass Cooling

- The cost of constructing a new house with significant thermal mass might exceed your ability to pay.

- Retrofitting a conventional building with thermal mass can get complicated and expensive; an improperly modified structure can pose a danger to its occupants.
- Some people consider thermal mass unattractive, no matter how well disguised. (On the other hand, some appreciate the ambience, saying that it "feels substantial.")
- In a home built into the poleward-facing slope of a hill, and with few or no windows that admit direct sunlight, the relative lack of natural light can pose a problem for some people.

PROBLEM 5-4

If I paint the exterior of my home white, or cover it with white siding, won't it impede the radiation of thermal energy away from the building at night, thereby reducing the effectiveness of the thermal mass in the exterior wall for cooling purposes?

SOLUTION

A white exterior surface does, in fact, slow down the radiation of thermal energy away from the structure at night. But this disadvantage is more than offset by the fact that a white exterior surface greatly retards the absorption of heat when the sun shines on it. In the hottest part of the year at temperate latitudes, daylight takes up a much greater part of the day than darkness.

Still Struggling

Perhaps someone will invent a paint that's normally dark or black, but that turns white when exposed to sunlight. If applied directly to the exposed, outside surface of a thermal mass that forms an exterior wall, this sort of paint would enhance the radiation of heat away from the building at night, while impeding the absorption of heat during the day!

Evaporative Cooling

We've learned that when water changes from a liquid to a gaseous state, it absorbs thermal energy. The water doesn't have to actively boil for this process to occur. The effect takes place at any temperature where water normally exists in liquid form (a range that extends far above and below the normal human comfort zone). By accelerating the process of water evaporation, we can make a body of liquid water effectively remove thermal energy from a home's interior air. This phenomenon gives us the principle behind *evaporative cooling*.

How It Works

When hot, dry air passes through a water-soaked foam or fiber *filter* several millimeters thick, some of the water evaporates. As a result, the air emerges from the filter a few degrees cooler than the air that came in. Reservoirs, which receive a constant supply of fresh (ideally distilled) liquid water, ensure that the filters don't dry out.

Figure 5-5 is a simplified functional diagram of an evaporative cooler designed for mounting on a flat, level rooftop. The assembly has a cubical shape,

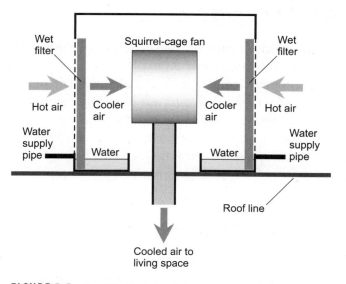

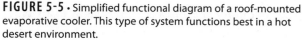

FIGURE 5-5 · Simplified functional diagram of a roof-mounted evaporative cooler. This type of system functions best in a hot desert environment.

and occupies roughly the same physical volume as a conventional air conditioner. The filters, designed for maximum water uptake by capillary action, absorb water from the reservoirs. A floating-ball device similar to the ones used in toilet tanks keeps the water level constant in the reservoir.

A powerful *squirrel-cage fan* pulls hot outside air in through the filters. The hot air causes some of the water in the filters to evaporate, so the air emerges from the filter at a temperature considerably below that of the outside air. The cooled, more humid air flows into the living space. The air outlet comprises a single vent in the center of the ceiling of the room chosen for cooling, producing convective currents in which cool air descends in the middle of the room and moves radially outward near the floor. Air leaves through vents near the floor, so you get a constant supply of fresh, cool air for the living space.

TIP   *The system shown in Fig. 5-5 works best in a single-level building. Ideally, you should locate it in the room that's used by the largest number of people, or that's occupied most often.*

? Still Struggling

An evaporative cooling system will function at peak efficiency in hot, dry weather, but it will work to some extent even if the outdoor temperature is only a little above the comfort level. The poorest performance occurs under conditions of high humidity. Evaporation can't take place very well when the air contains a lot of water vapor. Evaporative coolers enjoy the most popularity in hot desert regions, such as southern California, Arizona, and New Mexico. If you travel to the Australian Outback, you'll also encounter them.

Advantages of Evaporative Cooling

- An evaporative cooling system uses only about 1/5 of the electrical energy consumed by a conventional air conditioner to provide the same amount of cooling. Only the blower draws significant current from the electric utility.

- Because evaporative systems consume minimal electricity, they cause less greenhouse gas production or other pollution from the utility power plants.

- Evaporative cooling systems have simpler designs than conventional refrigeration type units, so fewer things can go wrong! If the unit does malfunction, it rarely costs much money to repair it.

- Evaporative cooling systems don't use ozone-depleting compounds, as some older refrigeration type cooling systems do.

- The living space receives a constant supply of outside air, eliminating the "stuffiness" or "clamminess" that can sometimes occur with recirculation in conventional air-conditioning systems.

- An evaporative system won't produce the annoying and unhealthy "meat freezer" effect, in which people entering a building from the outside on a hot day "hit a cold wall."

- Evaporative systems provide humidification in desert locations, mitigating physical complaints associated with extreme low humidity. A conventional system, in contrast, drives the moisture content of the air down, even when it's dry to begin with.

Limitations of Evaporative Cooling

- An evaporative cooling system won't work well if the humidity is high because the outdoor air entering the system is nearly saturated and can't accept much moisture from the filters.

- In a large building with many rooms, a single evaporative cooling unit can serve only one room or zone. Multiple units, one for each room or zone, along with insulated ducts, can provide some cooling in such situations, but this arrangement drives up the cost and increases the system complexity.

- An evaporative system may not provide enough cooling for comfort if the outdoor temperature becomes exceptionally high, as can occur in locations such as the Mojave desert in southern California during the summer.

- If an evaporative cooling system leaks water, damage can occur to a flat roof and the interior ceiling below if the roof doesn't have proper drainage.

- You must frequently clean or replace the filters, or they'll grow moldy, giving rise to health problems. If the outdoor environment contains a lot of particulate matter, you'll have to clean the filters even more often.

PROBLEM 5-5

My house has a steeply pitched roof. How can I install an evaporative cooler up there?

SOLUTION

You can build a level support platform at the peak of the roof, on which the cooling unit can rest. You can install a straight duct below the blower to let the air flow down into the living space through the attic. The platform structure must be strong enough to support the weight of the unit and the water it contains, and must be secured to the roof well enough so that a high wind won't rip the whole thing loose. The duct must have exterior insulation so that the cooled air doesn't reheat as it passes through the attic.

Subterranean Living

You can achieve the ultimate insulation from heat and cold by placing your entire living structure underground. The ceiling of the highest level will lie just below the surface, and your roof can serve as your yard or driveway. You might go even further, placing your *subterranean living* structures considerably beneath the surface. In the extreme, a few adventurous people have built homes in defunct nuclear-missile silos!

How It Works

The temperature beneath the earth's surface remains remarkably constant, once you get down to a certain depth (a few meters). This rule holds true even in locales where the difference between the highest and lowest air temperatures during a typical year can be extreme! In most temperate regions, the subterranean temperature at a depth of a few meters hovers around 10°C (50°F). As you go deeper than that, the temperature gradually rises.

Subterranean homes must use concrete and steel in their construction so that they can withstand the pressure that builds up beneath the earth's surface. You can't have windows that provide direct views to the outside, of course. However, you can install mirrors to build "periscope" type windows and skylights! You'll also have to install a sophisticated, reliable ventilation system to ensure a constant flow of fresh air to the interior.

TIP *A subterranean structure should incorporate some method of shock absorption to keep earth vibrations, caused by motorized traffic or other activity at the surface, from disturbing occupants in the living space.*

Still Struggling

As with any type of housing, subterranean living works better in some places than in others. Obviously, you won't want to build an underground home in an earthquake zone if you want to live to reach a ripe old age! You'll also want to avoid locations that have concentrations of radioactive minerals, volcanic activity, underground caverns, or underground streams. All such places are too dangerous for living beneath the earth's surface. In addition, a subterranean dwelling or structure must lie entirely above (and well above) the water table, regardless of the location.

Advantages of Subterranean Living

- Subterranean living can eliminate heating and cooling costs almost entirely.
- An underground house offers an absolute storm shelter. Even an extreme tornado won't directly affect the structure, although surface-mounted ventilation duct openings, solar panels, wind turbines, or other peripheral apparatus will likely suffer damage or destruction.
- Subterranean living is practical in congested areas, even in the downtowns of large cities. (This asset has been realized in a portion of Tokyo, Japan, in the form of a project called *Geotropolis*.)
- With proper acoustical design, an underground home can offer complete insulation from "noise pollution" originating at the surface. You'll never have to hear a screaming siren, barking dog, loud party, or street fight again!

Limitations of Subterranean Living

- Underground construction constitutes a prescription for disaster in an earthquake zone.

- Underground buildings do not last long in any location with expansive soil, or where earth movement occurs. You must check the history of a location, and scrutinize existing structures (especially older ones), for signs of settling or soil expansion before you consider building a subterranean home there.

- Subterranean living can induce *claustrophobia* (fear of confined spaces) in susceptible individuals.

- You can't expect to have panoramic views of the outdoors from an underground house.

- Radon gas can pose a problem unless the structure incorporates exceptional ventilation in conjunction with airtight interior floors, walls, and ceilings.

- A well-designed and meticulously maintained ventilation system is mandatory for any underground dwelling. Such systems can prove expensive.

- You can't build an underground dwelling in a location where the water table remains high, or in any zone prone to flooding.

PROBLEM 5 - 6

How can I get good natural light in a subterranean home without building "periscopes"? Wouldn't a massive opening at the surface, along with a huge array of mirrors, degrade the thermal insulation? I want to build an underground house, but I'm concerned about the negative health effects of daylight deprivation.

✔SOLUTION

If you don't want to install "periscope" type windows or skylights, you can buy specialized electric lamps that produce illumination resembling daylight. You can place these lamps at strategic locations throughout the home, turning them on at dawn and off at dusk. And of course, you can always spend a lot of time outdoors, doing all of the things that surface dwellers do, such as walking, hiking, cycling, and commuting.

QUIZ

Refer to the text in this chapter if necessary. A good score is eight correct. You'll find the correct answers listed in the back of the book.

1. In a sidehill-constructed home incorporating thermal mass for cooling in the Australian Outback, the side with the most window area, looking downslope, should ideally face toward the
   A. north.
   B. south.
   C. east.
   D. west.

2. To ensure that the heating elements operate as they normally would if powered by the utility, a direct wind-powered electric heating system should include a
   A. storage battery.
   B. solar panel.
   C. thermal mass.
   D. voltage regulator.

3. In a desert environment, an evaporative cooling system offers the secondary benefit of
   A. removing toxic gas from the indoor air.
   B. removing excess water vapor from the indoor air.
   C. improving a home's thermal insulation.
   D. humidifying the indoor air.

4. Theoretically, a residential hydroelectric power system produces a form of "pollution" by transferring small amounts of
   A. greenhouse gas to the atmosphere.
   B. thermal energy to the stream water.
   C. particulate matter to the stream water.
   D. toxic chemicals to the earth near the stream.

5. In most locations, the subterranean temperature
   A. drastically increases in summer and fall, and drastically decreases in winter and spring.
   B. gradually increases as you go deeper than a few meters beneath the surface.
   C. gradually decreases as you go deeper than a few meters beneath the surface.
   D. never exceeds the outdoor air temperature, no matter how cold the weather gets.

6. **You'd do well to consider building a subterranean home in a location**
   A. known for high concentrations of radioactive minerals because you can use the nuclear energy to gain independence from the commercial utilities.
   B. next to an underground stream because you can use it as a water source and also as a hydroelectric power source.
   C. near a volcano because you can use its limitless energy to operate a geothermal heat pump.
   D. other than any of the above; they're all dangerous places for underground living.

7. **With moderate wind, you can expect a midsize residential wind turbine, with all of its output energy dissipated in electric heating elements, to make those elements put out roughly**
   A. 3410 Btu/h.
   B. 12,000 Btu/h.
   C. 41,000 Btu/h.
   D. 341,000 Btu/h.

8. **In a fast-flowing small river with sufficient vertical drop, you can expect a well-engineered water turbine, with all of its output energy dissipated in electric heating elements, to make those elements put out roughly**
   A. 70,000 Btu/h.
   B. 3410 Btu/h.
   C. 2400 Btu/h.
   D. 1200 Btu/h.

9. **In bright sunlight, a set of six silicon PV cells in series will produce approximately**
   A. 0.5 V.
   B. 1.5 V.
   C. 3.0 V.
   D. 6.0 V.

10. **The maximum amount of electrical power that you can get from a solar panel in bright sunshine, assuming that the panel is made up of identical PV cells, depends on**
    A. the total surface area of the panel.
    B. the number of series-connected cells.
    C. the type of storage battery used.
    D. the number of appliances connected to the panel.

chapter 6

Conventional Propulsion

Conventional propulsion systems include machines powered by gasoline, diesel fuel, jet aircraft fuel, and rocket fuel. All of these fuels work by generating "controlled explosions" in which combustible substances yield *kinetic energy* along with chemical by-products. Gasoline- and diesel-powered vehicles are sometimes called *fossil-fuel vehicles* (FFVs).

CHAPTER OBJECTIVES

In this chapter, you will

- Review the properties of gasoline and petroleum diesel fuel.
- See how internal-combustion and diesel engines operate.
- Compare gasoline to petroleum diesel fuel for propulsion purposes.
- Learn how aircraft fuel derives from crude oil and kerosene.
- See how a jet engine works.
- Compare solid and liquid rocket fuels.

Gasoline Fuel and Engines

Automobiles, trucks, trains, boats, propeller-driven aircraft, and various other machines "burn" *gasoline* in *spark-ignition engines*, which are specialized *internal-combustion engines*. People in the United Kingdom call gasoline *petrol* (pronounced "PET-rel") because it comes from petroleum.

What Is Gasoline?

At room temperature and standard atmospheric pressure, gasoline is an aromatic, flammable liquid mixture of chemical compounds called *hydrocarbons*. We've seen some examples of such compounds: methane, heating oil, and propane. Gasoline bears a chemical relation to all of them. If we burn one U.S. gallon (1 gal) of gasoline outright, it yields approximately 1.14×10^5 Btu of thermal energy.

Gasoline has high *volatility*, meaning that it easily vaporizes under everyday conditions. At normal atmospheric pressure, gasoline vaporizes faster as the temperature rises. For this reason, refineries produce specific, individualized grades of gasoline for various geographical regions, and also for different seasons of the year in any specific region. The most volatile gasoline serves in winter or cold locations, and the least volatile gasoline serves in summer or warm locations.

Oil refineries produce gasoline along with other fossil fuels. Constituents of gasoline include compounds known as *alkanes*, *alkenes*, and *cycloalkanes* in proportions that depend on the nature of the crude oil available and the grade of gasoline produced. The highest grades of gasoline contain the highest amounts of *octane*, a hydrocarbon that helps prevent *engine knock*, which can take place when gasoline ignites too fast in an internal-combustion engine.

Besides hydrocarbon atoms, gasoline contains *additives* intended to improve the efficiency with which it burns. In the middle of the twentieth century, refiners added *tetraethyl lead* to gasoline because it, like octane, can minimize engine knock. By the 1980s, tetraethyl lead fell from favor as researchers learned of its toxicity. Lead from spent fuel accumulates in the environment, and therefore, in food products and water supplies. Nowadays, you'll hardly ever encounter "leaded fuel." Other additives vary from country to country.

TIP   *Some grades of gasoline contain approximately 10 percent added* **ethanol,** *also called* **ethyl alcohol,** *usually derived from corn or grain. Another fuel known* *as E85 comprises 85 percent ethanol and 15 percent gasoline. Most vehicles that burn ordinary gasoline can also burn fuel containing 10 percent ethanol, but it takes a specialized engine to properly burn E85.*

How a Gasoline Engine Works

A gasoline engine takes advantage of the flammable nature of gasoline to generate mechanical energy. The gasoline mixes with air in a *carburetor* to form a fine mist of gasoline droplets. This mixture passes into an enclosed *cylinder* containing a movable *piston* (Fig. 6-1). An *electric arc*, produced by a *spark plug*, ignites the gasoline/air mixture, causing a small explosion that makes the vapors in the cylinder violently expand to push the piston downward, a process called *ignition*. The piston connects to a rotatable *crankshaft* through a rod and bearings. The rotation of the crankshaft carries the rod around because the assembly has *angular momentum*. Therefore, the piston slides back upward in the cylinder after it has reached its lowest position. Each single upward or downward passage of the piston constitutes a *stroke*.

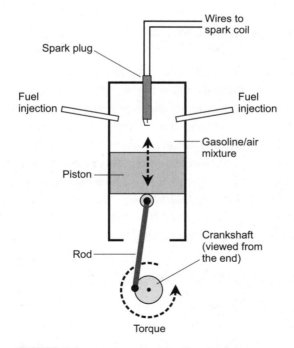

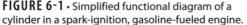

FIGURE 6-1 · Simplified functional diagram of a cylinder in a spark-ignition, gasoline-fueled engine.

In a *two-stroke engine*, ignition takes place every time the piston reaches the top of its cycle, so two piston strokes occur (one down, one up) for every single ignition event. In a *four-stroke engine*, ignition takes place at the top of every other complete cycle, so four piston strokes occur (down, up, down, up) for every ignition event. If the gasoline/air *injection* and the spark ignition occur repeatedly and rhythmically, and if everything happens in proper "sync" with everything else, the crankshaft rotates with enough *torque* to turn a *gear drive* or *belt drive* connected to a wheel, propeller, electric generator shaft, or other rotating assembly.

TIP   *The efficient operation of a gasoline-fueled engine depends on precise, constant timing. Mathematically, the engine efficiency is the ratio of the actual kinetic energy output to the total potential energy required to run the machine. Improper timing degrades the efficiency and increases the physical strain on the engine components. In the extreme, poor timing can cause total engine failure.*

?

Still Struggling

Most internal-combustion engines contain more than one cylinder of the type shown in Fig. 6-1. The strokes follow one another in a staggered sequence to provide smoother operation than a single cylinder can provide. Therefore, you'll often hear about *two-cylinder*, *four-cylinder*, *six-cylinder*, and *eight-cylinder* engines. Sometimes you'll encounter engines with more than eight cylinders. In general, as the number of cylinders increases, so does the mechanical energy that an internal-combustion engine of a given size and mass can produce. (There's a "point of diminishing returns," however. You should not expect to find anyone manufacturing a 128-cylinder engine!)

Advantages of Gasoline for Propulsion

- Gasoline has a high *energy-to-mass ratio*, also known as *energy density*. In other words, for a given quantity of fuel, you can get a lot of work. Few other fuels can rival gasoline in this respect.
- Gasoline engines provide a lot of mechanical power in proportion to their bulk and weight.

- A well-manufactured gasoline engine offers excellent ruggedness and reliability. If you take care of one, it can provide service for decades.
- A well-maintained gasoline engine can operate over a wide range of temperatures, humidity levels, and barometric pressure levels, encompassing almost all conditions encountered on this planet.

Limitations of Gasoline for Propulsion

- Gasoline combustion, even when complete, produces carbon dioxide (CO_2), a known greenhouse gas.
- When gasoline does not burn completely (as is the case in any real-world engine), some carbon monoxide gas (CO) is produced. This gas poses a deadly danger if it leaks into a vehicle, or if a gasoline engine runs in an enclosed area.
- Gasoline presents a fire hazard if it leaks from storage tanks or from any machine that contains it.
- Sulfur compounds in some forms of gasoline can contribute to a form of pollution called *acid precipitation*, in which water droplets in clouds become contaminated with sulfuric acid, spreading over large regions and returning to the earth as precipitation.
- Gasoline engines, when used in machines such as "leaf blowers" or "weed whackers," can cause annoying *noise pollution*.
- Some of the compounds in gasoline increase the risk of cancer in humans and animals directly exposed to them over a period of time.
- Gasoline derives from crude oil, the price of which can "spike" after geopolitical or natural disasters.
- Crude oil, the basis for gasoline and various related fuels, constitutes a non-renewable resource. When we use it all up, it'll be gone for good.

PROBLEM **6 - 1**

Haven't emission-control devices and regulations practically eliminated the pollution problems caused by gasoline engines?

SOLUTION

Emission reductions afforded by technology and legislation have been largely offset by the increased number of gasoline engines in use

throughout the world, especially in countries with rapidly growing economies. Emission-control technology doesn't get rid of all the pollutants produced by gasoline engines. The problem with CO_2 has gotten worse in
recent years. This gas doesn't pose a direct toxicity risk in humans, but it
gives rise to *greenhouse effect* in the atmosphere, reducing the natural
radiation of thermal energy into space. Many scientists believe that this
imbalance, sustained over a long period, will contribute to significant and
potentially troublesome climate change.

Petroleum Diesel Fuel and Engines

Conventional *diesel fuel*, also called *petroleum diesel* or *petrodiesel*, works well in
applications similar to those in which we find gasoline. Engineers prefer diesel-
fueled engines over gasoline-fueled engines to propel large trucks, locomotives,
agricultural implements, and other heavy vehicles.

What Is Diesel Fuel?

Diesel fuel comes from petroleum in the oil refining process. Like gasoline, diesel fuel is a hydrocarbon mixture. But diesel fuel has even greater energy density
than gasoline. If you set 1 gallon of petroleum diesel on fire, it will yield approximately 1.29×10^5 Btu of thermal energy by the time it's completely spent.

Because of the greater energy density of diesel fuel when compared with
gasoline, a diesel engine can travel farther on a given volume of fuel than a
gasoline-powered engine of the same size can, when pulling the same mass over
the same terrain. Diesel fuel also has roughly 18 percent greater physical density than gasoline. Some engineers consider diesel fuel to bear a closer resemblance to heating oil than to gasoline.

Some diesel fuels contain vegetable oils and/or liquefied animal fats along
with petroleum-derived hydrocarbons. The proportion of *biological oil* to petroleum fuel can vary over a wide range. Biologically derived diesel fuel is known
as *biodiesel*. The addition of biodiesel to petroleum diesel reduces the emission
of sulfur compounds when the fuel burns; the biological product contains less
sulfur than the petroleum product. This difference may prove significant if the
use of biodiesel becomes widespread because sulfur emissions have been impli-
cated as a cause of acid precipitation.

How a Diesel Engine Works

A diesel engine is a compression-ignition system that generates mechanical energy by means of combustion, just as a gasoline engine does. However, a diesel engine does not have spark plugs, so it doesn't need the supporting electrical system that spark plugs require.

Figure 6-2 illustrates the operation of a single cylinder in a diesel engine. The process resembles what happens in a gasoline-engine cylinder. As the piston moves upward, the heat produced by compression ignites the fuel/air mixture. When the piston nears or reaches the top of its up-and-down motion cycle, the fuel/air mixture passes into the cylinder so that combustion can take place.

TIP *Timing constitutes a critical factor for a diesel engine to operate with optimum efficiency and minimum pollutant emissions. Most diesel engines contain multiple cylinders, as do most gasoline engines. As the number of cylinders increases, so does the mechanical energy that the engine can produce (subject to the same practical limitation we see with spark-plug-type engines).*

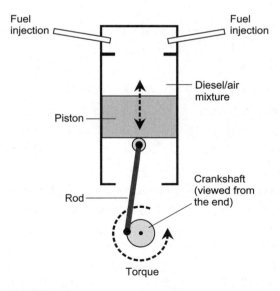

FIGURE 6-2 · Simplified functional diagram of a cylinder in a diesel-fueled engine. Note the absence of electrical components.

Advantages of Petroleum Diesel for Propulsion

- Diesel fuel has greater energy density than gasoline.
- In heavy vehicles, diesel fuel can deliver more power (actual mechanical energy per unit of time) than gasoline can.
- A diesel engine can offer better efficiency than a gasoline engine of the same size.
- The combustion of diesel fuel produces less deadly CO gas than the combustion of an equal amount of gasoline.
- Diesel engines are more reliable than gasoline engines because diesel engines have no electrical systems.

Limitations of Petroleum Diesel for Propulsion

- Petroleum diesel fuel contains more sulfur than gasoline, producing higher emissions of sulfur dioxide and other sulfur-containing pollutants.
- A diesel engine weighs more than a gasoline engine capable of producing the same maximum propulsion power.
- The exhaust from diesel fuel has an odor that some people find objectionable.
- When a diesel engine is not properly tuned, the exhaust contains soot, especially when the engine operates near full load. This soot contains unburned carbon, which can cause health problems for people in cities where large numbers of diesel engines operate.
- A diesel engine is harder to start and keep running than a gasoline engine, particularly in cold weather when diesel fuel thickens into a gel-like or even crystalline substance. When that happens, the fuel injector can't effectively deliver fuel into the cylinders.
- Some of the compounds in petroleum diesel fuel increase the risk of cancer in humans and animals directly exposed to them over a period of time.

PROBLEM 6-2

How can we overcome the cold-weather problems that occur with the use of diesel fuel?

✔️**SOLUTION**

In some vehicles, the fuel lines, fuel filter, engine block, and cylinders have built-in electric heaters to keep the fuel from thickening at low temperatures. When the vehicle runs, the engine's alternator can supply the electricity for these heaters. When the vehicle sits idle, an external utility power source can supply the electricity.

Conventional Jet Fuel and Engines

Propeller-driven aircraft, such as private and small commuter planes, burn high-octane gasoline. Most commercial and military aircraft employ jet propulsion, which requires a different sort of fuel. Some aircraft use turbine-driven propeller engines known as *turboprops*. Jet and turboprop engines can burn a mixture of high-octane gasoline and jet fuel.

What Is Jet Fuel?

Most jet fuel derives from a compound known as *kerosene*, which arises in the oil refining process along with gasoline, petroleum diesel, and other products. Kerosene can also come from coal. (As early as the mid-1800s, people used coal-derived kerosene as fuel for lamps that provided indoor and outdoor illumination.)

Kerosene has various non-aircraft uses. In Japan, some people use it for home heating. Small portable stoves, often used by campers and mountain climbers, also employ this fuel. Kerosene produces a spectacular flame when burned in open air, and has found favor in the entertainment industry for this reason. It works well as a general solvent. General-purpose kerosene is less refined than the compound that forms the basis of jet fuel. It has a characteristic odor, similar to that of diesel fuel, that can give some people nausea or headaches.

When refiners process kerosene for use as aircraft fuel, the sulfur content is reduced, as are its natural corrosive properties. The most common kerosene-derived jet fuel used in the United States is called *JET A*. It freezes at −40°C (−40°F) and has an *autoignition temperature* of approximately 425°C (800°F). Some other jet fuels, notably *JET B*, freeze at lower temperatures, but they have greater volatility, and find use only in high-altitude or polar flight where frigid temperatures prevail.

TIP *Kerosene-derived jet fuels contain additives, such as* antioxidants *(to keep the fuel from becoming too viscous),* electrostatic dissipation *substances (to prevent "static electricity" from causing sparks and consequent fires or explosions), chemicals to reduce the corrosive nature of pure kerosene, and* icing inhibitors *to prevent fuel-line freezing at the subzero temperatures commonly encountered at high altitudes.*

How a Jet Engine Works

Figure 6-3 is a functional (but not literal) illustration of a basic jet engine. The proportions are exaggerated in some cases and minimized in others to clarify the interaction among the components.

Air enters through a large opening. A *compressor*, also called a *fan*, drives the air into the engine. The compressed air passes down the length of the engine into the *combustion chamber*, where fuel injection takes place and the air-fuel combination burns. The drastic increase in temperature, caused by the combustion, produces extreme pressure in the chamber. Hot gases emerge from the rear of the combustion chamber at high speed. A turbine in the gas stream provides power for the compressor once the engine has begun to run. (Initially, an external power source must start up the compressor. In the earliest jet engines, power for the compressor came from an external piston engine similar to the ones used in propeller aircraft.) The exhaust leaves the rear of the engine

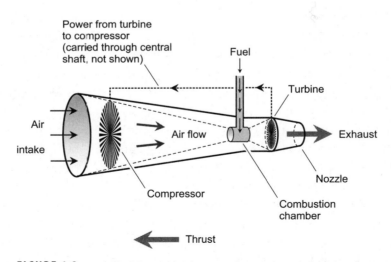

FIGURE 6-3 · Simplified functional diagram of a conventional jet aircraft engine.

through the *nozzle*. If the speed of the exhaust exceeds the forward airspeed of the whole assembly, *thrust* occurs.

TIP *Jet engines work best at high airspeeds (more than approximately (640 km/h or 400 mi/h) because they accelerate small amounts of air by a large factor. For airspeeds less than that, propeller-driven aircraft often perform better because they accelerate large amounts of air by a small factor. Turboprops also work well for aircraft at slower speeds. In recent years, turboprops have begun to supplant propeller engines for small and medium-sized commuter aircraft.*

Advantages of Conventional Jet Fuel

As of this writing, propeller, turboprop, and jet engines enjoy a single, but overwhelming, asset in aviation: No alternatives can compete with them! Engineers have begun looking into fuel cells, hydrogen, and even nuclear power as alternative methods for providing aircraft propulsion, but none of these technologies are anywhere near ready for mass implementation today.

Limitations of Conventional Jet Fuel

- The combustion of aircraft fuel, even when complete, produces CO_2.
- Aircraft fuel can create a fire hazard, if not properly stored.
- Flammable aircraft fuel can occasionally gain entry to aircraft cabins or otherwise leak from the tanks or fuel lines, creating the risk of fires and explosions.
- Sulfur compounds in some forms of aircraft fuel can contribute to acid precipitation.
- Conventional aircraft engines cause noise pollution. In recent years, low-noise jet engines have evolved, but for airports in urban areas, the noise can nevertheless be a nuisance.
- Jet engines often produce *contrails* (sometimes called "vapor trails") that might affect the dynamics of the atmosphere and climate by increasing the earth's percentage of high-level cloud cover over time.
- Some of the compounds in aircraft fuel increase the risk of cancer in humans and animals directly exposed to them over a period of time.
- Aircraft fuel derives from crude oil, so it suffers from all the market-related and supply-related problems that go along with that energy source. The only alternative is coal, not a very attractive option.

PROBLEM 6-3

Suppose that hydrogen becomes widely available someday, replacing natural gas for home heating, as well as gasoline and diesel for cars, trucks, boats, and trains. Won't it work well in aircraft, too?

✔ SOLUTION

Yes, but only if scientists and engineers can develop a cost-effective and safe way to obtain, transport, and store usable hydrogen fuel in the necessary amounts.

Conventional Rocket Fuel and Engines

Today's rocket fuel falls into two categories: *liquid rocket fuel* and *solid rocket fuel*. A liquid-fueled rocket engine works like a gigantic jet engine, except that the oxidant supply travels with the vehicle rather than being supplied by external air. A solid-fueled rocket engine differs from a jet engine in its basic design. In a sense, the rocket resembles a huge, controlled-combustion firecracker containing an oxidant mixed in with the powder.

What Is Liquid Rocket Fuel?

Liquid rocket fuel consists of a *propellant* and an *oxidizer*. The propellant in a rocket engine serves as the counterpart of the fuel in a jet engine, and the oxidizer in a rocket engine serves as the counterpart of the air that allows the fuel in a jet engine to burn.

Common rocket-engine propellants include kerosene, hydrogen (liquefied for storage in onboard tanks), and a nitrogen/hydrogen compound called *hydrazine* (N_2H_4). In the case of kerosene and hydrogen propellants, pure oxygen (liquefied for storage in onboard tanks) forms the oxidizer. It's sometimes symbolized with the acronym LOX (not to be confused with smoked salmon!). In the case of the hydrazine propellant, a nitrogen/oxygen compound called *nitrogen tetroxide* (N_2O_4) serves as the oxidizer.

The cleanest-burning liquid rocket fuel is hydrogen, which, when combined with oxygen, yields only energy and water vapor. When kerosene is refined for use as a rocket fuel, few impurities exist. However, some CO and CO_2 gases inevitably result from the combustion of this fuel because of the carbon atoms in its molecules. Hydrazine and nitrogen tetroxide yield considerable quantities of nitrogen when they react, but nitrogen poses no toxicity risk. In fact, it makes up nearly three-quarters of the composition of the earth's atmosphere at the surface.

What Is Solid Rocket Fuel?

The earliest solid rocket fuels resembled gunpowder, and found their primary applications in fireworks and weapons. Today, small hobby rockets, also known as model rockets, use solid fuel. A typical model rocket "engine" is a small cylinder of packed gunpowder-like material about the size of your index finger. A wire filament, heated by electric current from a battery, ignites the fuel, which subsequently burns for one or two seconds. The resulting thrust can propel a small rocket, about half a meter or 18 inches tall, to an altitude of several hundred meters (maybe 2000 feet).

Basic solid fuel contains a propellant, an oxidizer, and a *catalyst* that facilitates steady, reliable combustion following ignition. These fuel constituents originally exist in powdered form. They're mixed and packed uniformly to ensure an even, sustained, timed burn. A typical military solid-fueled rocket engine contains charcoal (carbon) as the fuel, potassium nitrate as the oxidizer, and sulfur as the catalyst. Rocket scientists and engineers call this chemical concoction *black powder*.

TIP *Alternative materials that can provide solid rocket fuel include sodium chlorate, potassium chlorate, powdered magnesium, and powdered aluminum. Because of their light-colored appearance, rocket experts refer to these substances as* **white powder.**

Still Struggling

Neither black nor white powder can provide enough energy to accelerate a vehicle to *orbital velocity*, a speed of about 29,000 km/h (18,000 mi/h). Advanced solid fuels can produce sufficient thrust in the *booster* (first stage) of a multistage rocket to allow subsequent liquid-fueled stages to put a payload into orbit, or even achieve *escape velocity* for interplanetary travel. The minimum speed required to escape the gravitational influence of the earth is approximately 40,000 km/h (25,000 mi/h).

How a Liquid-Fueled Rocket Engine Works

Figure 6-4 illustrates the principle behind a liquid-fueled rocket engine. The propellant and the oxidizer flow into the combustion chamber, where an initial ignition event starts the combustion process. As long as the propellant and the oxidizer both keep flowing into the chamber, combustion continues. The depressurization, along with the heating, converts the liquid propellant and oxidizer to a hot gaseous mixture.

Forward thrust results from the *action/reaction* principle: the hot gases, escaping from the nozzle, produce a powerful backward jet, causing the vehicle to accelerate forward. In the atmosphere, forward thrust requires that the escaping gases attain a speed at least equal to the forward speed of the vehicle. However, once the vehicle reaches the relative vacuum of outer space, this principle no longer applies. In space, any escaping gas, regardless of its speed, imparts a force on the vehicle that causes it to accelerate in a forward direction.

How a Solid-Fueled Rocket Engine Works

Figure 6-5 illustrates a solid-fueled rocket engine of the sort used in model rockets. (In larger vehicles, the same principle applies, although the engine geometry might differ somewhat.) Once ignition occurs, the fuel burns at a

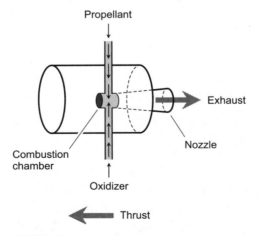

FIGURE 6-4 · Simplified functional diagram of a liquid-fueled rocket engine.

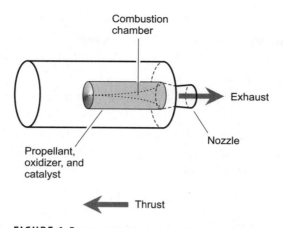

FIGURE 6-5 · Simplified functional diagram of a
solid-fueled rocket engine.

more or less constant rate, producing thrust as the hot-gas combustion by-products rush out of the rear opening or nozzle.

Once a solid-fueled engine ignites, combustion occurs until all the fuel is gone. You can't throttle such an engine down or stop the burn. For this reason, solid-fueled engines won't work for sustained, controlled rocket travel in outer space. Because solid fuel is customarily used only in the initial stages of a "rocket ship," serving to get the vehicle up to orbital or escape speed and not for navigation, this limitation poses no problem in practice.

Advantages of Conventional Rocket Fuel

Conventional rocket fuels—liquid or solid—provide the only practical methods for spacecraft propulsion at the time of this writing. Alternative technologies such as *ion engines*, controlled *nuclear fusion* reactions, and *solar sails* are in the conceptual stages, but these vehicle types will not likely come into common use until at least the end of the twenty-first century.

Limitations of Conventional Rocket Fuel

- With conventional rockets, you can't attain speeds much greater than the escape velocity from earth: 40,000 km/h or 25,000 mi/h. The huge amount of fuel required to produce higher speeds would make the vehicle too massive for acceleration to those speeds.

- Conventional rockets don't provide an efficient method of propulsion in an overall sense. Putting even a small payload into earth orbit requires an enormous amount of fuel (many times the mass of the payload itself).

- Conventional rockets won't work for interstellar travel because of their limited maximum speed. The nearest star besides the sun lies about 40 trillion (4.0×10^{13}) kilometers or 25 trillion (2.5×10^{13}) miles from our Solar System. Even at a speed of 1,000,000 kilometers per hour or 620,000 miles per hour, a journey to that star, *Proxima Centauri*, would take more than 4500 years!

PROBLEM 6-4

Can a conventional rocket ship employ a combination of liquid and solid materials to obtain its propulsion?

SOLUTION

Yes. This feat has been carried out in the form of *hybrid fuel* with a solid propellant and a fluid or gaseous oxidizer. However, the solid propellant, without the oxidizer mixed in uniformly, often burns unevenly or incompletely. Even so, *SpaceShipOne*, the first private manned rocket vessel, used hybrid fuel. The propellant comprised solid *hydroxy-terminated polybutadene* (HTPB), and the oxidizer comprised *nitrous oxide*.

QUIZ

Refer to the text in this chapter if necessary. A good score is eight correct. You'll find the correct answers listed in the back of the book.

1. Figure 6-6 is a simplified drawing of a cylinder in a petroleum-diesel-fueled engine. The fuel/air mixture passes into the cylinder when the top of the piston is nearest to
   A. point *X*.
   B. point *Y* on the way down.
   C. point *Y* on the way up.
   D. point *Z*.

2. What's the purpose of nitrogen tetroxide in a liquid-fueled rocket?
   A. It lowers the freezing point.
   B. It increases the energy density.
   C. It prevents explosions.
   D. It makes the fuel burn.

3. Tetraethyl lead was once added to gasoline fuel for use in motor vehicles because it helped to prevent
   A. engine overheating.
   B. gas-line freeze.
   C. engine knocking.
   D. vapor locks.

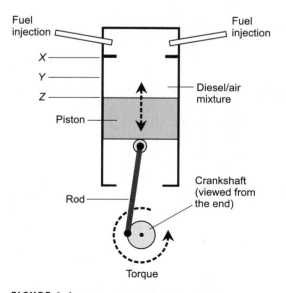

FIGURE 6-6 · Illustration for Quiz Question 1.

4. **Kerosene has been used in all of the following applications** *except*
   A. heating residential houses.
   B. as a solvent for cleaning.
   C. extinguishing accidental fires in entertainment displays.
   D. portable stoves for use when camping.

5. **What's the advantage of an electrostatic dissipation additive in jet aircraft fuel?**
   A. It lowers the boiling point.
   B. It raises the freezing temperature.
   C. It reduces the risk of explosion.
   D. It increases the energy density.

6. **In a four-stroke gasoline engine, ignition takes place at**
   A. the top of every piston cycle.
   B. the top of every other piston cycle.
   C. the top and bottom of every piston cycle.
   D. the top, bottom, and middle of every piston cycle.

7. **When a substance has high volatility, then by definition it easily changes state from**
   A. liquid to vapor.
   B. liquid to solid.
   C. solid to liquid.
   D. vapor to solid.

8. **Which of the following characteristics** *does not* **apply to petroleum diesel fuel?**
   A. It has greater energy density than gasoline has.
   B. It has greater physical density than gasoline has.
   C. It contains more sulfur than gasoline does.
   D. It can't crystallize at low temperatures as gasoline does.

9. **Solid-fueled rocket engines won't work well for long-distance space travel because**
   A. the fuel won't burn in a vacuum.
   B. the burn can't be stopped once it has started.
   C. the volatility is too low for use in a vacuum.
   D. it explodes when ignited in a vacuum.

10. **Conventional jet fuel and engines can contribute to the production of**
    A. carbon-dioxide.
    B. acid rain.
    C. contrails.
    D. All of the above

chapter 7

Propulsion with Methane, Propane, and Biofuels

Automotive engineers have tested many different combustible liquids and gases as possible substitutes for gasoline and petroleum diesel. A few such substances propel specialized motor vehicles today.

CHAPTER OBJECTIVES

In this chapter, you will

- Compare earth-derived methane with methane from biological sources.
- Contrast liquefied natural gas with compressed natural gas.
- Define and quantify gasoline-gallon equivalent (GGE) values.
- Compare grades of ethanol-gasoline blends.
- Learn how engineers make biodiesel fuel from old cooking grease.

Methane for Propulsion

Many people use methane as the fuel of choice for interior heating systems, especially in the United States. Methane can also serve as an alternative to gasoline or diesel fuel in motor vehicles.

How It Works

Methane-fueled cars existed before gasoline became widely available. However, more powerful, affordable, and mass-produced fossil-fuel vehicles superseded methane-fueled vehicles early in the twentieth century. Methane fuel for cars, trucks, and boats can come from the earth, but other methods have been developed for obtaining methane for use in motor vehicles—notably fermentation or composting of plant and animal waste. Such fuel goes by the name *biogas* or *biomethane*.

In small and medium-sized internal-combustion engines, methane fuel produces less carbon dioxide (CO_2) gas than gasoline does, in order to do the same amount of mechanical work. Methane combustion also produces considerably less smoke and particulate matter than the combustion of other fuels, particularly diesel, in small and medium-sized engines. In larger engines, these advantages are lost to some extent, and are further offset by marginal efficiency.

For motor-vehicle use, methane can condense at low temperatures to form so-called *liquefied natural gas* (LNG), greatly increasing the energy density compared with *compressed natural gas* (CNG) and mitigating one of the chief obstacles to application. As of this writing, a limited number of CNG and LNG refueling stations exist. The lack of a widespread infrastructure poses a problem for individuals and businesses interested in taking advantage of these fuels.

TIP   *In 2010, British engineers introduced a small passenger car called the Bio-Bug, powered by CNG derived from biological sources, and intended to serve as an alternative to electric cars. The Bio-Bug can burn conventional gasoline when the CNG runs out, and can attain speeds of up to 114 miles per hour (183 kilometers per hour) on a level surface under windless conditions, traveling 5.3 miles (8.5 kilometers) per cubic meter of decompressed methane.*

TIP   *Some individuals and small businesses have proprietary composting plants, and use the methane thereby obtained to power their own vehicles and heat their indoor environments.*

? Still Struggling

In recent years, engineers have developed new methods for getting "fossil-fuel type" methane gas out of the earth. A process called *fracking* involves pumping water down into the earth to force gas out. Methane can also be derived from coal by means of a process called *gasification*. Enter these terms into your favorite Internet search engine to learn about the latest developments involving them. Fracking, in particular, has grown controversial in recent years. Its detractors claim that it can cause contamination of drinking water supplies. Its proponents say that it will reduce American dependency on foreign fuel sources. Who is right—or are they both right? You decide!

Advantages of Methane for Propulsion

- Well-designed methane-fueled engines produce less CO_2, smoke, and particulate pollution than gasoline- or diesel-fueled engines in converting the potential energy into usable mechanical energy.

- The widespread production and use of methane may reduce the severity and impact of problems, such as variable prices and supply interruptions, that occur with gasoline and petroleum diesel fuel.

- Increased production of biogas for use in vehicles would, as a side benefit, increase the methane supply available for heating homes and businesses and for running electric power plants.

- In many countries, including the United States, methane distribution pipelines already exist.

- Methane can be produced in small-scale composting plants. The supply does not have to come exclusively from centralized sources. Diversified production could enhance civil security against natural disasters or political upheavals.

Limitations of Methane for Propulsion

- In its combustible form, and at ordinary room temperature and pressure, methane exists in the gaseous state. This presents handling and transportation problems, as with any industrial gas.

- While small and medium-sized methane-powered engines have reasonable efficiency, they don't offer the high performance obtainable in advanced gasoline-powered and diesel-fueled engines.

- Methane is not widely available at vehicle refueling stations in most countries at the time of this writing, even though the pipeline infrastructure does exist.

- The production of biogas by composting can produce objectionable odors and, if improperly carried out, can breed disease-causing microorganisms.

- Tanks that hold methane gas require periodic inspection and certification by licensed and qualified personnel, at considerable expense to the owner.

PROBLEM 7-1

Draw a flowchart that outlines the production, distribution, and use of biogas from animal and plant waste.

SOLUTION

Figure 7-1 illustrates a typical biogas production scheme suitable for use on a local or regional scale.

Propane for Propulsion

Propane, also known as *liquefied petroleum gas* (LPG), occurs as a by-product of petroleum refining. Propane remains popular for home heating and cooking, particularly in rural parts of the United States. Propane can also fuel medium-sized electric generators, such as those used by people who own recreational vehicles (RVs) or who have homes operating independently from the conventional utility grid. As with methane, propane was used to run vehicles before gasoline was available. Today, thousands of propane-powered cars and trucks exist in the United States.

How It Works

The main difference between propane and methane lies in the fact that propane, in storage, exists as a liquid rather than as a gas. Propane becomes gaseous

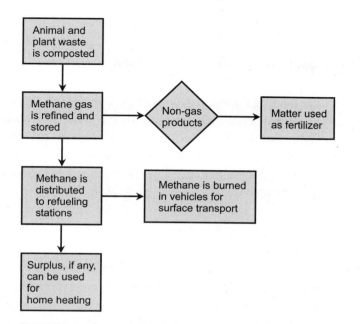

FIGURE 7-1 · Illustration for Problem 7-1 and its solution.

when released from its tank. Propane contains approximately 8.43×10^4 British thermal units per U.S. gallon (Btu/gal) of energy content.

Engineers sometimes speak of the *gasoline-gallon equivalent* (GGE) of an alternative liquid fuel, which they define as the ratio of the number of British thermal units available in a U.S. gallon of gasoline to the number of British thermal units available in a U.S. gallon of the alternative fuel. We can calculate the GGE of propane (let's call it G_p) as

$$G_p = (1.14 \times 10^5)/(8.43 \times 10^4)$$
$$= 1.35$$

In Chap. 6, we learned that one U.S. gallon of gasoline provides approximately 1.14×10^5 Btu when burned outright, which equals 1.35 times the energy available from the same volume of liquid propane. The real-world value can vary somewhat, depending on the grades of fuel and on their purity. The economic ramifications of any GGE figure depend on the price of the alternative fuel, the price of gasoline, the efficiency of a gasoline-powered engine, and the efficiency of an equivalent alternative-fuel-powered engine.

? Still Struggling

We can calculate the GGE figure using units other than the U.S. gallon and the British thermal unit, but only if we use the same units for both fuels being compared. For example, we can use liters and joules to calculate the GGE of propane, and as long as we use these units to define the energy per unit volume of both propane and gasoline, the resulting GGE figure will still turn out to be 1.35. The GGE is a *dimensionless quantity*: a ratio of two parameters that are both expressed in identical units.

TIP   *It takes about the same amount of time to fill the tank of a propane-powered vehicle as it does to fill a tank of the same size in a conventional gasoline- or diesel-fueled vehicle. Most light-duty propane-powered cars, trucks, and buses in use today have been converted from gasoline or diesel power. However, a few vendors offer new propane-powered vehicles. Thousands of propane refueling stations exist in the United States, but they are not as easy to find as regular gasoline or diesel refueling stations.*

Advantages of Propane for Propulsion

- A properly operating propane-fueled engine produces less polluting emissions than a properly operating gasoline- or diesel-powered engine of the same size. Such pollution includes ozone-forming compounds, as well as toxic chemicals, such as benzene, formaldehyde, and acetaldehyde.

- In some locations or at certain times, the cost of propane, considering its GGE, is lower than the cost of gasoline or diesel fuel, translating into a lower fuel cost overall for propane-powered vehicles as compared with other fossil-fuel vehicles.

- Although an extensive infrastructure for propane refueling does not yet exist, the cost of developing one would be lower than the cost of developing a similar infrastructure for certain other alternative energy technologies such as hydrogen.

- Most of the propane used in the United States comes from domestic sources, so events overseas have less effect on the supply.

Limitations of Propane for Propulsion

- You'll need special training to safely refuel, operate, and maintain propane-powered vehicles. Although propane poses less of an explosion risk than fuels that remain in gaseous form when stored, propane fuel does becomes a gas when it enters the atmosphere.

- The maximum operating range of a dedicated propane-powered vehicle is only about three-quarters of the maximum operating range of a gasoline-powered vehicle with the same size tank. (This figure comes from the ratio $1/G_p = 1/1.35 = 0.741$ or roughly 3/4.)

- Fuel tanks for propane must have more rugged construction than tanks used for gasoline or diesel fuel, resulting in greater vehicle mass and an adverse effect on fuel mileage and acceleration characteristics. The greater mass of a propane-powered vehicle also increases the required braking distance.

- Tanks that hold propane require periodic inspection and certification by licensed and qualified personnel. This process can prove inconvenient and costly.

PROBLEM 7-2

Are hybrid vehicles available that can run on either conventional fuel or propane? Can a gasoline- or diesel-powered vehicle be converted to run on propane?

SOLUTION

Yes, and yes again! A propane-powered vehicle can have a hybrid fueling system so it can run on either propane or conventional fuel. A gasoline or diesel fueling system can be modified into a hybrid or propane-only fueling system.

Ethanol for Propulsion

Ethanol derives from the fermentation of plant matter, so it's considered a *biofuel*. The ethanol production process resembles liquor manufacturing. As a fuel, ethanol is as old as methane and propane. Henry Ford favored the burning of ethanol in his Model T and other early cars. Ironically, he called ethanol "the fuel of the future." In its pure form, ethanol exists as a flammable, volatile, clear liquid.

WARNING! *Ethanol is the same stuff that makes you drunk when you overindulge in "alcoholic" beverages. But don't you dare drink industrial ethanol! It may, and often does, contain deadly additives or impurities.*

How It Works

In the United States, corn constitutes the crop of choice for ethanol production. In Brazil, they use sugar cane. In Canada, they use wheat. Theoretically, the *cellulose* from any plant matter, terrestrial or aquatic (even trees or seaweed), can serve as the basis for ethanol production. Figure 7-2 is a simplified flowchart of the *dry mill process*, a common method of obtaining ethanol from corn.

Most modern vehicles can burn gasoline with up to 10 percent ethanol added with no adverse effects. Blends that contain 10 percent or less of ethanol are sometimes called *E10* or *gasohol*. If the fuel contains more than 10 percent ethanol mixed in with gasoline, vehicle engines require modification to burn it efficiently. The percentage of ethanol in some blends is 85 percent, in which case the fuel is known as *E85*.

So-called *flex-fuel vehicles* can burn a variety of ethanol-gasoline mixtures. A sensor, connected to the computer that controls the engine, detects the percentage of ethanol in the fuel tank and adjusts the engine accordingly. The mileage per gallon with ethanol blends is a few percent lower than the mileage with pure gasoline because ethanol has a lower energy density than gasoline. The extent of this effect depends on the proportion of fuels in the blend; the more

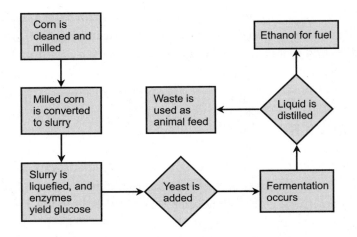

FIGURE 7-2 • The dry mill process makes it possible to derive ethanol from corn, and yields animal feed as a byproduct.

ethanol, the worse the mileage. But because added ethanol increases the octane rating of gasoline, performance often improves with gasohol or E85, provided that the vehicle has been modified (if necessary) to properly burn fuels with the added ethanol.

TIP *Failure to modify an engine for ethanol-containing gasoline, when necessary, can lead to poor mileage, poor performance, reduced engine life, and permanent damage to engine components.*

TIP *During the first years of the twenty-first century when the price of gasoline in the United States soared, interest grew in E85. Most cars at that time were not equipped to burn E85, and its distribution was limited. Most outlets appeared in the midwestern part of the United States, where corn is grown in abundance. By the time you read this, E85 might be more popular, its availability might be more widespread, and more cars and trucks might be capable of running on it. (On the other hand, maybe the opposite will happen, and E85 might be out of favor!) In any case, you should always check your vehicle manufacturer's specifications before attempting to burn E85 in your engine. If in doubt, consult a qualified dealer's mechanical experts.*

? Still Struggling

Crop farming for ethanol production takes up CO_2, a known greenhouse gas, from the atmosphere. That same CO_2 returns to the atmosphere when the ethanol is burned. Therefore, ethanol use is *CO_2 neutral* when produced from plants grown for that purpose. However, if plants are cut down to produce ethanol and the demised biomass is not replaced, the use of the resulting fuel causes a net increase in atmospheric CO_2.

Advantages of Ethanol for Propulsion

- Ethanol can serve as an extender and octane enhancer in conventional gasoline.
- Ethanol can function as a primary fuel (in E85), thus reducing dependence on petroleum products.

- Ethanol can reduce emissions of deadly CO gas.
- Ethanol production and use won't contribute to global CO_2 when responsibly done.
- Ethanol can help prevent gas-line freeze in extremely cold weather.
- In some locations, gasohol and E85 are cheaper per gallon than conventional gasoline.
- The widespread production and use of ethanol as a fuel benefits farmers by increasing the demand for their products.
- The plant matter used to produce ethanol constitutes a renewable resource.
- Ethanol is not as flammable as gasoline, and therefore, presents less of an explosion hazard.

Limitations of Ethanol for Propulsion

- Fueling stations for gasoline containing ethanol, and especially for E85, are not (at the time of this writing) as abundant as fueling stations that offer only conventional gasoline.
- In some locations, gasohol and E85 cost more per gallon than conventional gasoline.
- Inappropriate use of ethanol fuel can damage some vehicle engines, especially if E85 is inadvertently pumped into a vehicle designed to burn conventional gasoline.
- Because much of the ethanol produced today comes from plant matter that could otherwise serve as human food, some people argue that the widespread use of ethanol as fuel could contribute indirectly to world hunger.

PROBLEM 7-3

Can you use the ethanol from the fuel refining process to mix intoxicating drinks?

✔ SOLUTION

You had better not try to use ethanol fuel to "spike" drinks, hoping to get drunk off the mix! The law requires that a small amount of gasoline or other objectionable additive be mixed in with ethanol produced for use as fuel. If you consume this "pure ethanol," it will make you violently ill. It could even cause organ failure or death. Get your "drinking alcohol" at legitimate liquor stores, or in bars!

Biodiesel Fuel for Propulsion

Biodiesel is a combustible, viscous liquid consisting of alkyl esters of fatty acids derived from vegetable oil or cooking grease. This fuel burns well in some compression-ignition engines designed for petroleum diesel (and poorly in others).

How It Works

Soybeans constitute the cheapest and most abundant vegetable source of oil that can be refined to obtain biodiesel fuel. The extracted oil is processed to remove all traces of water, dirt, and other contaminants. Free fatty acids are also removed. A combination of methyl alcohol and a catalyst, usually sodium hydroxide or potassium hydroxide, breaks the oil molecules apart in a chemical reaction known as *esterification*. The resulting compounds, called *esters*, are then refined into usable liquid biodiesel fuel. Figure 7-3 is a simplified flowchart showing this process.

Used-up cooking oil and animal fats, which would otherwise constitute waste matter, can be refined to produce biodiesel fuel. The process resembles the way biodiesel derives from soybean oil, but includes additional steps, as shown in the flowchart of Fig. 7-4. Methyl alcohol and sulfur are used in a process called *dilute acid esterification* to obtain a substance resembling fresh vegetable oil, which is then processed in the same way as soybean oil to obtain the final product.

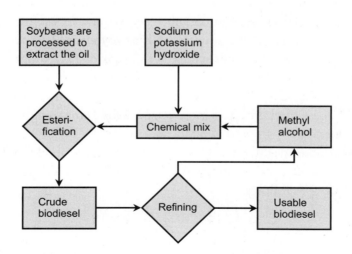

FIGURE 7-3 · Production of biodiesel fuel from soybeans.

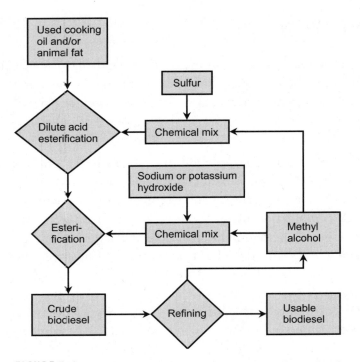

FIGURE 7-4 · Production of biodiesel fuel from used cooking oil and animal fat.

The combustion of biodiesel fuel produces less CO_2 gas than the combustion of petroleum diesel fuel. In addition, biodiesel contains less sulfur, particularly when derived from vegetable sources such as soybeans. As a result, we see reduced emissions of compounds such as sulfur dioxide (SO_2) that can contribute to environmental damage in the form of acid precipitation. Most other emissions are reduced as well, including deadly CO gas and particulate matter. However, biodiesel fuel produces more nitrous oxide emission than petroleum diesel fuel does.

When burned outright, pure (100-percent) biodiesel fuel yields approximately 1.18×10^5 Btu/gal of energy. We can quantify the GGE of biodiesel fuel (let's call it G_{bd}) as follows:

$$G_{bd} = (1.14 \times 10^5)/(1.18 \times 10^5)$$
$$= 0.97$$

TIP *Biodiesel fuel is commonly blended with petroleum diesel fuel. Vendors write the percentage of biodiesel after an uppercase letter B to denote the proportion. For example, a mixture of 20 percent biodiesel and 80 percent petroleum diesel yields B20. A mixture of equal parts biodiesel and petroleum diesel yields B50. Pure (or "neat") biodiesel is B100. Nearly all conventional diesel engines can burn blends from pure petroleum diesel fuel up to B20 without modification. In most diesel engines built since 1994, blends from B20 to B100 can be used with minor modification, but transportation and storage of these fuels requires special attention.*

? Still Struggling

Diesel engine owners should review their warranty statements before attempting to burn any substitute for pure petroleum diesel. Some manufacturers will void the warranty if an engine is operated on fuel that contains more than a certain percentage of biodiesel.

Advantages of Biodiesel for Propulsion

- Biodiesel combustion produces fewer emissions (with the exception of nitrous oxides) than the combustion of an equal amount of petroleum diesel.
- Biodiesel comes from renewable resources. We can literally grow the supply! In contrast, petroleum diesel comes from nonrenewable resources.
- Defunct cooking oils and grease can be used to produce biodiesel.
- The widespread use of biodiesel can reduce dependency on imported oil.
- Pure biodiesel is not toxic if spilled because it's biodegradable. Petroleum diesel can cause environmental damage if spilled in large amounts.
- Biodiesel is safer than petroleum diesel because it does not spontaneously burn as easily when stored or transported.

Limitations of Biodiesel for Propulsion

- At the time of this writing, biodiesel was not as widely available as petroleum diesel. Where biodiesel was available, it generally cost more per gallon than petroleum diesel. (Conditions may change by the time you read this book.)

- Storage, handling, and transportation of biodiesel requires special management.

- The combustion of biodiesel produces more nitrous oxide emissions than the combustion of an equal amount of petroleum diesel.

- Because some biodiesel is produced from soybeans that are a good source of protein as well as essential food oils, some people have expressed concern that the widespread use of biodiesel as fuel might contribute indirectly to world hunger.

- Biodiesel has solvent properties that can cause problems in older diesel engines. It can loosen deposits, clogging fuel filters. It can also damage rubber components.

- In most applications, we will observe a slight reduction in performance and mileage per gallon with biodiesel as compared with petroleum diesel.

PROBLEM 7-4

Can used cooking oil or grease, such as that from a deep-fat fryer or a frying pan, be poured into the fuel tank of a diesel-powered vehicle and directly consumed as biodiesel? For example, if the cook at a restaurant has a liter of bacon grease left over from the preparation of the day's breakfasts for his customers, can he pour the hot grease straight into the tank of his diesel truck and expect the grease to function as biodiesel?

✔SOLUTION

Absolutely not! Demised cooking oil or grease must be processed as shown in Fig. 7-4 before it's used as biodiesel fuel. This fact should be obvious in the case of bacon grease, which solidifies near room temperature. But it holds true even for plant fats that remain liquid at relatively low temperatures, such as corn oil, canola oil, or soybean oil.

QUIZ

Refer to the text in this chapter if necessary. A good score is eight correct. You'll find the correct answers listed in the back of the book.

1. **As a fuel for motor vehicles, methane offers all of the following advantages over gasoline and diesel fuel** *except*
   A. ease of small-scale production.
   B. improved engine performance.
   C. lower carbon-dioxide emission.
   D. reduced production of smoke.

2. **If we want to burn various gasoline/ethanol blends, we'll get the best results with**
   A. a conventional gasoline engine.
   B. a flex-fuel engine.
   C. an engine designed to burn pure ethanol.
   D. Any of the above; it doesn't matter.

3. **We can calculate the gasoline-gallon equivalent (GGE) for an alternative fuel by figuring out**
   A. the number of cubic meters of alternative fuel required to produce the same amount of energy as one liter of gasoline.
   B. the ratio of the energy available in a liter of the alternative fuel to the energy available (in the same units) in a liter of gasoline.
   C. the number of liters of the alternative fuel required to produce the same amount of energy as one liter of gasoline.
   D. the number of liters of gasoline required to produce the same amount of energy as one cubic meter of the alternative fuel.

4. **Fill in the blank to make the following statement true: "Most cars in use today can burn gasoline with up to _____ ethanol added without trouble."**
   A. 85 percent
   B. 50 percent
   C. 33 percent
   D. 10 percent

5. **Why is biodiesel fuel safer than petroleum diesel fuel, in general?**
   A. Biodiesel won't congeal at low temperatures during transport.
   B. Biodiesel can be more easily compressed for transport.
   C. Biodiesel won't start on fire as easily during transport.
   D. The premise of this question is wrong! Biodiesel is actually more dangerous than petroleum diesel.

6. One of the chief objections, in general, to the use of plant matter for producing fuel derives from the argument that
   A. plant matter constitutes a nonrenewable resource.
   B. using plant matter for fuel cuts into the world's food supply.
   C. plant-based fuels are prone to spontaneous combustion.
   D. plant-based fuels are difficult to transport.

7. Ethanol fuel for use in motor vehicles comes mainly from processed
   A. plant matter.
   B. methane.
   C. gasoline.
   D. cooking oil.

8. Biodiesel fuel for motor vehicles can be derived from processed
   A. methane.
   B. propane.
   C. petroleum diesel.
   D. cooking oil.

9. Fill in the blank to make the following statement true: "A gasoline-propelled vehicle with a full 16-gallon fuel tank can travel approximately _____ as a propane-propelled vehicle of the same weight with a full 16-gallon fuel tank before the need for refueling arises."
   A. the same distance
   B. 1.35 times as far
   C. 75 percent as far
   D. half as far

10. Biomethane generally results from a controlled process of
    A. composting.
    B. deep-earth mining.
    C. coal gasification.
    D. fracking.

chapter 8

Propulsion with Electricity, Hydrogen, and Fuel Cells

Many experts believe that the supply of petroleum-based products (so-called fossil fuels) will fail to keep pace with skyrocketing worldwide demand in coming decades, increasing the need for alternative motor-vehicle energy sources. Electricity offers a partial solution. So do *hydrogen* (the most abundant element in the universe) and specialized components known as *fuel cells*.

CHAPTER OBJECTIVES

In this chapter, you will

- Compare various battery technologies' assets and limitations.
- Learn how electric vehicles work.
- Compare series and parallel hybrid vehicle designs.
- Learn why hydrogen might someday serve as the ideal fuel for vehicles.
- Discover how fuel cells work, and how they can power electric vehicles.

Electric Vehicles

The basic concept behind the *electric vehicle* (EV) is straightforward. *Direct-current* (DC) electricity powers a large electric motor, which in turn propels the vehicle. The energy source usually comprises a rechargeable battery in the vehicle.

How They Work

In an EV, the motor connects to the wheels through a *drive system* similar to the *transmission* in an ordinary car or truck. The speed of the vehicle depends on the speed at which the motor runs. An accelerator pedal (or "gas pedal") controls the motor speed in an electric car or truck. A handlebar device controls the motor speed in an electric motorcycle or all-terrain vehicle (ATV).

The actual speed of the vehicle depends not only on the motor speed, but also on the gear ratio between the motor and the wheels. The motor gets its power from a rechargeable battery.

Most compact electric cars can travel about 100 km (60 mi) on a full charge before the batteries need recharging. The actual distance depends on the nature of the terrain, whether the vehicle travels with the wind or against it, and the temperature of the surrounding air. A good model can attain speeds up to about 130 km/h (approximately 80 mi/h) under ideal operating conditions: flat terrain, no wind, and moderate temperature.

Lead-Acid Cells and Batteries

Figure 8-1 is a functional diagram of a *lead-acid electrochemical cell*. A plate molded from *elemental lead* serves as the negative electrode, and a plate made of *lead dioxide* serves as the positive electrode. Both electrodes are immersed in a sulfuric-acid solution called the *electrolyte*. With this arrangement, a DC *voltage* develops between the electrodes. This voltage can drive an electric *current* through a *load*, such as an electric lamp, radio, computer, or motor. The *maximum deliverable current* depends on the mass and volume of the cell.

A battery consists of two or more identical cells connected together. In a battery made from identical cells connected in *series* (negative-to-positive, like the links in a chain), the total voltage increases directly in proportion to the number of cells, but the maximum deliverable current remains the same as that of any individual cell by itself. In a battery made from identical cells connected in *parallel* (negative-to-negative and positive-to-positive, like the rungs in a

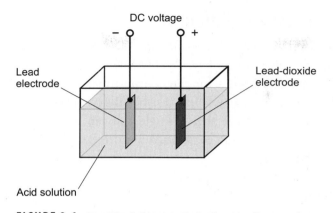

DC voltage

Lead electrode

Lead-dioxide electrode

Acid solution

FIGURE 8-1 · Simplified diagram of a lead-acid cell.

ladder), the total voltage equals that of any cell taken alone, but the maximum deliverable current increases directly in proportion to the number of cells.

If we connect a lead-acid battery to a load for a long time, the current gradually decreases, and the cell electrodes become coated with compounds as they react with the electrolyte. The nature of the acid electrolyte also changes. Eventually, all the chemical energy contained in the acid gets converted into electrical energy. Then the current drops to zero, and a potential difference no longer exists between the electrodes in the cells. However, if we drive current through the cells by connecting an external DC voltage source to the battery terminals for a period of time (negative-to-negative and positive-to-positive), the battery will regain its electrochemical energy, making it useful once more. That's the *recharging* process. We can repeat the *charge-discharge cycle* many times before a lead-acid cell or battery "wears out for good."

WARNING! *When recharging a lead-acid cell or battery, make sure that the positive wire from the charger goes to the positive terminal of the battery, and the negative wire from the charger goes to the negative terminal of the battery. If you get the terminals reversed, you'll in effect short out both the charger and the cell or battery. Under such conditions, some lead-acid cells and batteries will rupture or explode!*

TIP *Lead-acid batteries find application in most motor vehicles to provide the power for the initial startup ignition. You'll also find these batteries in* uninterruptible power supplies *for computer workstations, in wood-pellet and corn stoves to keep the blowers running in the event of a utility power failure, and in* notebook computers *as the main power source during portable operation.*

Nickel-Based Cells and Batteries

Nickel-based cells include the *nickel-cadmium* (NICAD or NiCd) type, shown in Fig. 8-2A, and the *nickel-metal-hydride* (NiMH) type, shown at Fig. 8-2B. The two types are identical except for the composition of the negative electrode. *Nickel-based batteries* are available in packs of cells connected together to obtain higher current and/or voltage than a single cell can provide. All nickel-based cells are rechargeable. They can go through hundreds of charge/discharge cycles if properly used and cared for.

You'll find nickel-based cells commercially available in various sizes and shapes. *Cylindrical cells* look like ordinary dry cells. (They're the types diagrammed in Fig. 8-2.) *Button cells* are small, pellet- or pill-shaped components often used in cameras, watches, memory backup applications, and other places

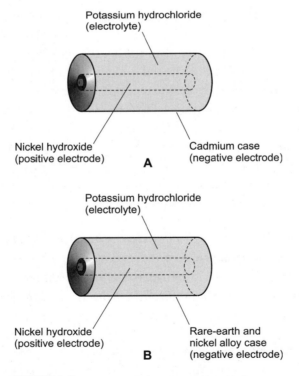

Potassium hydrochloride
(electrolyte)

Nickel hydroxide
(positive electrode)

Cadmium case
(negative electrode)

A

Potassium hydrochloride
(electrolyte)

Nickel hydroxide
(positive electrode)

Rare-earth and
nickel alloy case
(negative electrode)

B

FIGURE 8-2 · At A, simplified diagram of a nickel-cadmium (NICAD) cell. At B, simplified diagram of nickel-metal-hydride (NiMH) cell.

requiring miniaturization. *Flooded cells* are used in heavy-duty situations requiring a lot of energy for a long time, so they lend themselves to use in EVs. *Spacecraft cells* reside in sealed packages that can withstand the rigors of a deep-space environment.

A discharge anomaly peculiar to NICAD cells and batteries has become known among engineers as *memory effect* or *memory drain*. If you use a NICAD unit over and over, and if you discharge it to the same extent every time you use it, the component will lose much of its current-delivering capacity at that point in its discharge cycle. This phenomenon creates the impression that the battery has "died." You can sometimes resolve a memory problem by using the cell or battery *almost* all the way up (but not quite completely), fully recharging it, and then repeating the process several times.

TIP *Recently developed NiMH cells and batteries don't suffer as often from memory drain as NICADs do. In addition, a typical NiMH cell can store up to 40 percent more energy than a NICAD cell of the same mass and volume.*

? **Still Struggling**

Never discharge a nickel-based cell or battery all the way until it "dies." If you make this mistake, the polarity of one or more of the cells might reverse. Once a polarity reversal occurs, you'll find the battery useless, and you'll have to get rid of it. But don't throw it into an ordinary trash container! Take it to an authorized battery recycling center.

Lithium Cells and Batteries

Lithium cells gained popularity in the early 1980s, and have become increasingly prevalent since then. We can find several variations in the chemical makeup of these cells. They all contain lithium, a light, highly reactive metal. These cells, like all other cells, can be stacked to make batteries.

Lithium batteries originally found application as *memory backup* power supplies for electronic microcomputers. Lithium cells and batteries can last for years in very-low-current applications, such as memory backup or the powering of a digital *liquid-crystal-display* (LCD) watch or clock. Today, lithium batteries can be found in notebook computers, digital tablet devices, e-book (electronic-book) readers, cell phones, and various other portable and mobile devices.

Lithium-based cells and batteries can store a great deal of energy in a component of modest volume and weight. For this reason, many engineers believe that lithium-based batteries will someday play a key role in electric motorized personal transportation, particularly in smaller vehicles such as motorcycles, snowmobiles, and compact cars.

Storage Capacity

Any cell or battery has a certain amount of electrical energy that we can specify in *watt-hours* or *kilowatt-hours*. One watt-hour (1 Wh) represents the equivalent of one watt of electrical power (1 W) expended for one hour (1 h) of time. One kilowatt-hour (1kWh) equals 1000 Wh. Some engineers quantify a battery's storage capacity in terms of the mathematical *integral* of deliverable current with respect to time, in units of *ampere-hours* (Ah). The energy capacity in watt-hours equals the ampere-hour rating multiplied by the battery voltage.

A 1000-Ah battery can deliver 100 A for 10 h, or 10 A for 100 h, or 1 A for 1000 h. As you can imagine, there exist infinitely many current/time combinations, and almost any of them (except for the extremes) can be used in real life. The extreme situations are the *shelf life* and the *maximum deliverable current*. We define the shelf life as the length of time that the battery will remain usable if we never connect it to anything and just let it "sit on the shelf." We define the maximum deliverable current as the highest amount of electrical current that the battery can drive through a load without having its voltage drop significantly because of *internal resistance*, and without internal overheating.

Battery Charging

We can recharge storage batteries in a variety of ways. The most common method involves the use of an external charging unit connected to a source of conventional utility power such as a wall outlet. A charging system using a wind turbine, water turbine, or solar panel can also work. Solar panels can be placed on a house roof, vehicle roof, vehicle trunk lid, or vehicle hood to provide supplemental charging when the vehicle sits idle in the sun. In more advanced vehicles, energy for recharging can come from a primary internal combustion engine and an *alternator*, a set of *fuel cells*, or a special braking system that slows the vehicle down by "borrowing" kinetic energy from the wheels and using it to charge the battery.

The Pollution Question

At first thought, we might suppose that EVs don't generate any pollution whatsoever, but indirectly they do. For one thing, lead-acid and NICAD batteries can damage the environment if carelessly discarded, and unfortunately, a lot of people make that mistake. More significant, however, is the fact that the energy needed to charge an automotive battery has to come from somewhere, and that's usually an electric power plant that burns some sort of fossil fuel.

The sulfuric acid in a lead-acid battery produces fumes including sulfur dioxide, a known pollutant gas. Hydrogen, a flammable gas that can explode if confined and exposed to flame or spark, is also produced. Lead and cadmium are heavy metals, and constitute known environmental toxins. Therefore, you must take special precautions when discarding old lead-acid or NICAD batteries. Because of pollution concerns, NiMH batteries have replaced NICAD types in many applications. In most practical scenarios, a NiMH battery can directly replace a NICAD battery of comparable voltage and current-delivering capacity, and the powered-up device will work the same.

When you charge up the battery in an EV, you'll need a little more energy than you'll actually get from the battery over the course of its discharge period. For example, a 12-V battery rated at 2000 Ah will require a little more than 24,000 Wh, or 24 kWh, of energy to acquire a full charge. That's the equivalent of a portable electric space heater running on the "high" setting for 16 hours. This amount of energy will run a compact electric car in the city for approximately 80 km (50 mi) under ideal conditions.

If the number of EVs in use were to increase, a corresponding increase in the demand for electricity from power plants would result. Many of these power plants use methane, oil, or coal to run the generators. Even so, scientists generally agree that if EV usage displaces conventional vehicle usage kilometer-for-kilometer without any increase in the total number of kilometers driven by the population, the *overall* amount of motor-vehicle-generated pollution will diminish.

TIP   *According to some estimates, the overall pollution generated by EVs, kilometer-for-kilometer, is only about 10 percent of the overall pollution generated by fossil-fuel vehicles.*

Advantages of EVs

- The use of EVs can help industrialized countries reduce their dependency on foreign sources of oil.

- The overall pollution generated by EVs, kilometer for kilometer, is a fraction of that generated by vehicles that use combustion engines, even when we take the pollution from utility plants and battery manufacture into consideration.

- The cost of the energy required to operate an EV, kilometer-for-kilometer, is lower than the cost of the energy required to operate a fossil-fuel vehicle.

- In some locations, people who buy and use alternative vehicles, including EVs, can get tax breaks or rebates.

- With a little bit of ingenuity on the part of the owner, EV batteries can derive some or all of their charge from sources other than the electric utility.

- An EV can provide its owner with a sense of independence.

Limitations of EVs

- The maximum operating range for an EV, starting with a fully charged battery, is less than the operating range of a typical fossil-fuel type vehicle.

- The battery in an EV can lose some of its ability to hold a charge or deliver sufficient current when the temperature falls far below zero Celsius (the freezing point of water).

- An EV might not provide enough power to operate reliably in severe weather, particularly in heavy snow.

- The EV design concept does not lend itself very well to heavy-duty applications such as hauling freight or plowing snow.

- Because most EVs have small physical size and mass, drivers must use extra caution to keep themselves and their passengers safe while sharing the road with larger, heavier vehicles.

- In some areas, EV servicing can prove difficult or impossible because of a lack of parts or competent technicians.

PROBLEM **8-1**

How can you keep the interior of an EV at a comfortable temperature when the outdoor air reaches high or low extremes?

✔ SOLUTION

Modern EVs employ air-source heat pumps to maintain habitable interior temperature. You can preheat or precool the interior while the EV remains connected to the charging station. Enhanced thermal insulation can also help to keep passengers comfortable. But in extreme conditions, particularly when the outdoor air drops to dangerously low levels, maintaining a comfortable interior temperature while driving will drastically reduce the operating range between charges.

Hybrid Electric Vehicles

Electric and conventional energy sources "combine forces" in the *hybrid electric vehicle* (HEV). Two basic types of HEV exist today: the *series design* and the *parallel design*.

Series HEVs

In a series HEV, the mechanical propulsion comes from an electric motor powered by a rechargeable battery. In this sense, the series HEV resembles a simple EV. But instead of relying on an external source of energy for battery charging, the series HEV carries its charging system onboard as a generator driven by a small internal-combustion engine. The engine and generator run continuously while the vehicle operates. The engine can get its power from gasoline, methane, propane, E85, petroleum diesel, biodiesel, or hydrogen.

Figure 8-3A shows the general configuration of a series HEV. The generator includes a charging unit in the form of an alternating-current-to-direct-current (AC/DC) converter. The energy storage medium comprises a large rechargeable battery of the same type used in an EV. An optional external charger can be used, when the vehicle is not running, to keep the battery "topped off." The electric motor connects mechanically to a transmission that resembles the drive system in an EV.

TIP *A series HEV always runs from the electric motor, never from an internal-combustion engine. For this reason, this type of vehicle has limited speed and acceleration capabilities. Electric motors can't provide the short-burst high performance that an internal-combustion engine can produce. The primary asset of the series design is the fact that the limited combustion ensures low fuel usage and minimal polluting emission.*

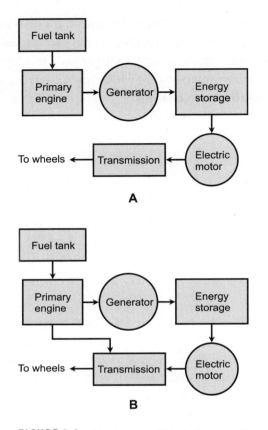

FIGURE 8-3 · At A, functional block diagram of an HEV with series design. At B, functional block diagram of an HEV with parallel design.

Parallel HEVs

The outstanding feature of a parallel-design HEV is the fact that both the primary engine and the electric motor contribute directly to propulsion. The two systems operate in tandem (parallel), with the burden automatically shifting from one to the other, depending on driving conditions from moment to moment. The primary engine usually gets its power from conventional gasoline or gasohol, although some primary engines use alternative fuels such as methane or propane.

Figure 8-3B shows the interconnection of components in a parallel HEV. As in the series design, the generator includes an AC/DC converter to charge the battery. An external charger can be used when the vehicle is parked. The electric motor and primary engine both connect to the transmission. At low speed, or when relatively little power is required to propel the vehicle, the electric motor predominates. At high speed, or under conditions of high power demand (when climbing a steep hill or passing another vehicle, for example), the internal combustion engine delivers.

? Still Struggling

In a parallel HEV, a microcomputer determines which engine should provide more power. The microcomputer is programmed to shift the relative burden between the two engines based on the amount of *back pressure* (mechanical resistance) in the drive system. As the back pressure increases, indicating a demand for high performance or speed, more of the burden falls to the internal combustion engine. As the back pressure decreases, such as it does on flat or downgrade terrain at moderate speed, more of the burden goes to the electric motor.

Regenerative Braking

In EVs and HEVs, the battery can derive extra charging energy by means of a scheme called *regenerative braking* when the vehicle decelerates (slows down) or travels down a steep grade. A generator and AC/DC converter, connected into the drive chain, harness some of the energy that would otherwise go to waste heating the brake drums. When the wheels turn against the high mechanical resistance of a generator connected to a heavy load, the effect is to *decelerate* the vehicle (slow it down), just as conventional brakes would do. Some engineers call this phenomenon *dynamic braking*.

In an EV or HEV equipped with regenerative braking, a microcomputer determines when the situation demands that power should come from the wheels rather than go to them. The wheels then supply the mechanical torque for the battery-charging generator, reducing the amount of energy that would otherwise be necessary to charge the battery from the primary engine or the external charger. This process not only saves energy and improves overall mileage, but it also reduces wear on the brakes, increasing the length of time between costly brake service appointments.

TIP   *Heavy trucks and train locomotives can take advantage of the deceleration caused by a generator connected to the wheels. If you ever see a sign in mountainous terrain that says "Dynamic Braking Prohibited," you know that residents in the area don't want to hear the noise that the process can create when a large engine and drive system do it.*

Advantages of HEVs

- The use of HEVs can help industrialized countries reduce their dependency on foreign sources of oil.
- The overall pollution generated by HEVs, kilometer for kilometer, is less than the overall pollution generated by vehicles that use combustion engines exclusively, even when we take the emissions from utility plants and battery manufacture into consideration.
- The cost of the energy required to operate an HEV, kilometer-for-kilometer, may be less than the cost of the energy required to operate a fossil-fuel vehicle (after the up-front investment).
- In some locations, people who use alternative vehicles, including EVs, can qualify for tax breaks or rebates.
- An HEV can operate from gasoline alone, just like a conventional fossil-fueled vehicle, if the battery discharges in the middle of a trip.
- The parallel HEV concept can serve in the design of heavy-duty vehicles, such as buses and large trucks. Many such vehicles have adopted this technology.

Limitations of HEVs

- A typical HEV costs more, in terms of the purchase price, than a conventional vehicle of the same size.
- Because of the complexity of their design, it sometimes proves difficult to find competent technicians who can service HEVs. This problem occurs most often in rural areas, or in cities where people rarely use HEVs.
- Many of the components in an HEV are specialized or proprietary, and are less available in remote locations than parts for fossil-fueled vehicles.
- In extremely cold weather, the electrolytes in an HEV's battery can freeze, causing battery failure.

 **PROBLEM 8-2**

Does any method exist for storing electrical energy derived from braking or downhill coasting, without involving the battery, if an intense burst of power is called for a short while later?

✔ **SOLUTION**

Yes. A device called an *ultracapacitor* or *supercapacitor* can accomplish this task. The component operates on the principle of separation of charge, in the same way as the capacitors work in DC power supplies designed for electronic equipment. An ultracapacitor can deliver a large current for a short time to provide an intense burst of mechanical power from the electric motor. These devices are employed in some sophisticated HEVs.

Hydrogen-Fueled Vehicles

When burned in the presence of pure oxygen, hydrogen (the lightest and most abundant element in the universe) liberates only energy and water vapor. In the atmosphere, which contains approximately 71 percent nitrogen, hydrogen combustion produces small amounts of nitrous oxide gas as well. Engineers abbreviate the term *hydrogen-fueled vehicle* as HFV.

How to Obtain Pure Hydrogen

Hydrogen does not naturally exist on earth in its free state; we always find it bound up in chemical compounds, most notably water. We can extract hydrogen gas from methane by a process known as *steam reforming*. In steam reforming, methane reacts with steam (hot water vapor) in the presence of elemental nickel, which acts as a catalyst. In addition to the hydrogen, carbon monoxide gas comes off as a by-product.

We can also get pure hydrogen gas by *electrolysis of water*. A molecule of pure water contains two atoms of hydrogen and one atom of oxygen chemically bound together. This bond can break when an electric current passes through liquid water. Electrolysis requires the addition of a mineral *electrolyte*, such as salt, sodium bicarbonate, or sulfuric acid, to enhance the water's ability to conduct electricity.

When we impose a significant DC voltage between two electrodes immersed in the water/electrolyte solution, hydrogen bubbles appear at the negative electrode, while oxygen bubbles appear at the positive electrode (Fig. 8-4). We can collect and store the gas from these bubbles. No dangerous by-products result. We must make sure to use DC, not AC such as that from household utility outlets. (If we apply AC rather than DC to the electrodes, hydrogen and oxygen gases appear together at both electrodes, and we can't separate them.)

Electrolysis is an expensive and inefficient process. It will take half again as much (50 percent more) energy to produce a cubic meter of hydrogen at standard atmospheric pressure by means of water electrolysis, as we will get when we burn that same cubic meter of hydrogen. In order to produce hydrogen on a large scale by means of electrolysis, some engineers have suggested the construction of huge dedicated solar- or wind-powered "electrolysis farms." It doesn't take much water to produce a lot of hydrogen, and saline or alkali lakes exist in various places that already contain the necessary electrolyte minerals.

TIP   *Alternative methods of electrolysis can break water molecules apart to obtain hydrogen. These methods involve the use of certain chemicals and/or heat. Another way to get hydrogen is to break down coal or biomatter (substances largely consisting of* **carbohydrates**) *into their elemental components*

How to Store and Transport Hydrogen

On-board fuel storage presents a technical problem that engineers will have to solve before hydrogen becomes practical for use as a motor-vehicle fuel. We can

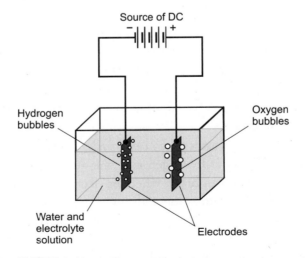

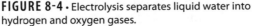

FIGURE 8-4 · Electrolysis separates liquid water into hydrogen and oxygen gases.

compress hydrogen gas in metal tanks, just as we can do with other gases. However, if we expect to provide enough fuel to propel a hydrogen vehicle for a reasonable distance between refueling stops, we'll need a tank so large and massive that it impairs vehicle efficiency and safety. In addition to this conundrum, compressed-gas tanks can pose a danger to life and property. In the event of a collision that causes the neck of the tank to break, the resulting release of pressure can turn the tank into a deadly missile.

Other fuel storage methods are under consideration. One technology involves the use of *metal hydrides* (compounds of metal and hydrogen), from which the hydrogen can be liberated at a controlled rate under certain conditions. Another technology makes use of a phenomenon called *gas-on-solid adsorption* (not *absorption*!). In the adsorption process, hydrogen gets "locked up" in complexes of microscopic carbon structures called *nanotubes*. Hydrogen gas can condense inside these structures at density levels comparable with the best storage tanks, but without the dangers of high pressure associated with tanks. Controlled liberation is possible, and the resulting hydrogen gas can serve directly as a fuel source. Both of these technologies, as of this writing, remain in the research-and-development phase.

TIP *Engineers have adapted conventional fossil-fuel vehicles so that they'll run on hydrogen, just as conventional vehicles have been modified to run on methane or propane.*

Advantages of HFVs

- Hydrogen, when burned with pure oxygen, does not produce any hazardous pollution whatsoever.

- Hydrogen engines offer better efficiency than methane or propane engines in converting the energy from the fuel into mechanical energy. In fact, hydrogen engines compare favorably with gasoline engines in this respect.

- The widespread production and use of hydrogen may mitigate problems, such as variable prices and supply interruptions, that occur with conventional fuels.

- Increased production of hydrogen for use in vehicles would, as a side benefit, increase the hydrogen supply available for heating homes and businesses.

- In many countries, including the United States, distribution pipelines for methane gas already exist. Some of these pipelines might be adapted to hydrogen for use in a network of refueling stations.
- Entrepreneurs can produce hydrogen in small-scale and local facilities, enhancing the security of the civilized world by distributing energy resources and assets.

Limitations of HFVs

- In its combustible form, and at ordinary room temperature and pressure, hydrogen exists as a gas, posing storage, handling, and transportation problems. In particular, designing a safe fuel tank for a hydrogen-powered vehicle is a challenge.
- As of this writing, motorists will rarely find hydrogen at vehicle refueling stations in most countries, including the United States.
- Tanks that hold hydrogen gas require periodic inspection and certification by licensed and qualified personnel.
- As of this writing, hydrogen remains expensive, mainly because of the cost of the processes involved in separating it out from naturally occurring compounds such as methane or water.

PROBLEM 8-3

How far can a motorist expect to drive an HFV on the amount of hydrogen derived, by electrolysis, from a liter of water?

SOLUTION

An efficient HFV gets about the same mileage from the hydrogen contained in a liter of water as a comparable fossil-fuel vehicle gets from a liter of gasoline.

Fuel-Cell Vehicles

Hydrogen, methane, propane, and other fuels burn directly to obtain propulsion, as we've seen. However, these fuels (and practically any other) can indirectly propel a specialized form of EV known as a *fuel-cell vehicle* (FCV).

What's a Fuel Cell, Anyway?

In the late part of the twentieth century, a new type of electrochemical power device emerged that holds promise as an alternative energy source: the *fuel cell*. The most talked-about fuel cell during the early years of research and development became known as the *hydrogen fuel cell*. As its name implies, it derives electricity from hydrogen. The hydrogen combines with oxygen (it *oxidizes*) to form energy and water, along with a small amount of nitrous oxide if air serves as the oxidizer. When a hydrogen fuel cell "runs out of juice," a new supply of hydrogen will get it working again.

Instead of literally burning, the hydrogen in a fuel cell oxidizes in a controlled fashion, and at a much lower temperature. Several schemes exist for making this process go smoothly. The *proton exchange membrane* (PEM) *fuel cell* represents one of the most widely used technologies. A PEM hydrogen fuel cell generates approximately 0.7 V DC, a little less than half the voltage of a typical electrochemical dry cell. To get higher voltages, individual cells are connected in series, so that their voltages add up. For example, to obtain 14 V DC, we would connect 20 hydrogen fuel cells in series because $20 \times 0.7 \text{ V} = 14 \text{ V}$. A series-connected set of fuel cells technically forms a battery, but engineers and technicians more often use the term *stack*.

Increased current-delivering capacity can be obtained by connecting cells or stacks in parallel, so that the current-delivering capacities of the individual cells or stacks add up. (The voltage of a parallel-connected set of identical cells or stacks equals the voltage of any single cell or stack all by itself.) For example, if we connect five stacks in parallel, each rated at 14 V DC and capable of delivering up to 10 A, the resulting combination will provide 14 V DC at up to 50 A because $5 \times 10 \text{ A} = 50 \text{ A}$.

Fuel-cell stacks can be obtained in various sizes from commercial vendors. A stack about the size and weight of a suitcase full of books can power a subcompact electric car. Smaller cells, called *micro fuel cells*, can provide electricity for devices that have historically operated from conventional cells and batteries, such as portable radios, lanterns, and notebook computers.

A fuel cell can get its "juice" from sources other than hydrogen. Almost any liquid or gas that will combine with oxygen to generate energy has aroused interest among engineers. *Methanol*, a form of alcohol, has proven easier to transport and store than hydrogen because it exists as a liquid at room temperature. Propane and methane have been used to provide the energy for fuel cells. Even gasoline, petroleum diesel fuel, and biodiesel fuel can work!

TIP *Some "alternative energy purists" object to the use of propane and methane in fuel cells because these substances (except biomethane) derive from fossil-fuel sources.*

How FCVs Work

An FCV essentially constitutes an EV that uses a fuel cell in place of, or in addition to, a storage battery. Figure 8-5 shows the functional basics of a hydrogen FCV. The electric motor connects to the wheels through a drive system similar to the transmissions found in fossil-fuel vehicles. A hydrogen FCV embodies some of the assets of EVs, HEVs, and HFVs all together.

In an FCV, the motor gets its power from the electricity provided by the fuel cell, and also, in some designs, from a rechargeable storage battery that can derive its charge from the fuel cell and also from a regenerative braking system of the type used in advanced EVs and HEVs. The use of a storage battery offers a special advantage: If the fuel tank goes empty, the battery can provide some extra driving range.

A typical FCV can convert about 50 percent of the energy contained in the hydrogen gas into usable electrical energy. The remainder of the energy ends up as heat, which gets dissipated into the atmosphere through a cooling system that resembles the radiator and associated apparatus in a fossil-fuel vehicle.

TIP *As with nearly all motor vehicles, the speed of an FCV depends on the speed at which the motor runs, and also on the gear ratio between the motor and the wheels.*

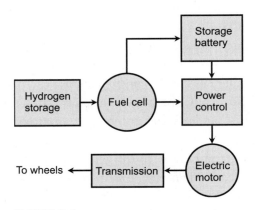

FIGURE 8-5 · Functional block diagram of a hydrogen FCV.

Still Struggling

Because hydrogen burns cleaner than fossil fuels, we can reasonably expect that a well-designed hydrogen FCV will operate more efficiently than a traditional fossil-fuel vehicle, although perhaps at a lower level of efficiency than a well-designed HFV that burns hydrogen directly to obtain propulsion.

Advantages of FCVs

- The use of FCVs can help industrialized countries reduce their dependency on foreign sources of oil.
- Increased production of hydrogen for use in FCVs would, as a side benefit, increase the hydrogen supply available for heating homes and businesses.
- Existing methane pipelines could be adapted for use in a network of FCV refueling stations, using either methane or hydrogen as the fuel.
- The overall pollution generated by hydrogen FCVs, kilometer for kilometer, is a fraction of that generated by fossil-fuel vehicles, even when we consider the pollution from hydrogen production and battery manufacture.
- Small-scale and local facilities can produce hydrogen.
- In some locations, people who use alternative vehicles, including FCVs, can qualify for tax breaks or rebates.

Limitations of FCVs

- The maximum operating range for a hydrogen FCV, starting with a full fuel tank, is less than the operating range of a conventional fossil-fuel vehicle. If other types of fuel cells (methane or propane, for example) are used, the operating ranges of FCVs and conventional fossil-fuel vehicles are comparable.
- In some areas, it can prove difficult to get FCVs serviced because of a lack of parts or competent technicians.
- The onboard storage and transport of fuel for hydrogen FCVs presents a major technological obstacle to the widespread deployment of these vehicles.

- As of this writing, hydrogen FCVs remain expensive to operate, largely because of the cost of the processes involved in separating hydrogen from naturally occurring compounds.

PROBLEM 8 - 4

If hydrogen gas can burn directly as fuel for a combustion engine in an HFV, what's the point in converting the energy from the hydrogen into electrical energy to run a motor in a hydrogen FCV? Doesn't that process introduce an unnecessary step that actually reduces the efficiency of an otherwise workable system?

✔ SOLUTION

A hydrogen FCV offers at least two advantages over a vehicle that burns hydrogen directly.

1. A fuel cell operates at a lower temperature than a hydrogen combustion engine, so that the process is inherently safer.
2. A hydrogen FCV can employ an on-board storage battery, from which the electric motor can directly obtain energy, if necessary. The battery also allows for storage of energy derived from regenerative braking.

QUIZ

Refer to the text in this chapter if necessary. A good score is eight correct. You'll find the correct answers listed in the back of the book.

1. How many PEM fuel cells would we need to place in series to get a battery that produces 2.8 volts?
   A. Two
   B. Four
   C. Six
   D. Eight

2. Regenerative braking can improve the operating range of an EV or HEV (between necessary fueling or recharging breaks) by
   A. reducing the friction in the brake drums.
   B. reducing the friction between the tires and the road.
   C. using the battery's energy to help brake the vehicle.
   D. using deceleration energy to charge the battery.

3. How can we recharge a lead-acid electrochemical cell?
   A. We can connect an external AC voltage source to the battery terminals for awhile.
   B. We can connect an external DC voltage source to the battery terminals for awhile, negative-to-negative and positive-to-positive.
   C. We can connect an external DC voltage source to the battery terminals for awhile, negative-to-positive and positive-to-negative.
   D. We can't!

4. If we burn a mixture of pure hydrogen gas and pure oxygen gas in the optimum proportion (two hydrogen atoms for every oxygen atom), we get
   A. energy and nothing else.
   B. energy and water vapor, but nothing else.
   C. energy, water vapor, and carbon monoxide.
   D. energy, water vapor, and particulate matter.

5. In theory, we can obtain pure hydrogen gas by
   A. electrolyzing water.
   B. steam reforming of methane gas.
   C. breaking down carbohydrate substances.
   D. All of the above

6. Memory drain in a NICAD battery can cause it to
   A. produce too much voltage.
   B. generate alternating current (AC) instead of direct current (DC).
   C. give you the false impression that it has "died."
   D. catch on fire, rupture, or even explode.

7. In an FCV, the fuel cell takes the place of the
   A. generator in an HEV.
   B. alternator in a gasoline-fueled vehicle.
   C. storage battery in an EV.
   D. Any of the above

8. In a series HEV, the mechanical power to provide the propulsion comes from
   A. an electric motor at all times.
   B. an electric motor when high performance is not needed, and a gasoline-fueled engine when high performance is needed.
   C. an electric motor at all times, supplemented by a gasoline-fueled engine when high performance is needed.
   D. a gasoline-fueled engine at all times, supplemented by an electric motor when high performance is needed.

9. In an EV, what methodology, device, or system helps to keep the interior reasonably warm in cold weather?
   A. A heat pump
   B. Residual engine heat
   C. An electric space heater
   D. Redirected road friction

10. In a battery comprising two or more identical electrochemical cells connected in series,
   A. the maximum deliverable current and the total voltage remain the same as that of any one of the cells alone.
   B. the maximum deliverable current and the total voltage both depend on the number of cells.
   C. the maximum deliverable current depends on the number of cells, and the total voltage remains the same as that of any one of the cells alone.
   D. the total voltage depends on the number of cells, and the maximum deliverable current remains the same as that of any one of the cells alone.

chapter **9**

Advanced Propulsion Methods

In this chapter, we'll look at propulsion alternatives that will prove interesting in the near-term and medium-term future for three modes of transportation: trains, ships, and spacecraft. For trains, *magnetic levitation* holds promise. For ships, *nuclear power* may make a resurgence. For spacecraft propulsion, *ion rockets*, *hydrogen-fusion engines*, and *solar sails* have been considered.

CHAPTER OBJECTIVES

In this chapter, you will

- See how magnetic fields can act as antigravity forces.
- Learn why magnetic levitation sometimes works and sometimes doesn't.
- Compare the assets and limitations of magnetic-levitation trains.
- Learn how nuclear reactions can propel oceangoing vessels.
- Discover alternatives to conventional rockets for long-distance space travel.

Magnetic Levitation

Magnetic levitation takes advantage of magnetic forces to suspend moving objects above fixed media. The technology works because objects having strong magnetic poles of the same sense (that is, north-and-north or south-and-south) exhibit a mutual, powerful repulsive force when they are brought into close proximity.

Earnshaw's Theorem

Imagine approximately 100 small, pellet-shaped permanent magnets glued down, evenly spaced, on the inside surface of a plastic bowl with the north poles facing upward. This arrangement forms a large magnet with a concave, north-pole surface. Suppose that you anchor this bowl to a tabletop, and then take a single pellet-shaped magnet and hold it with its north pole facing downward over the center of the bowl, as shown in Fig. 9-1A. As soon as you release the single magnet, it flips over and sticks to one of the magnets inside the bowl.

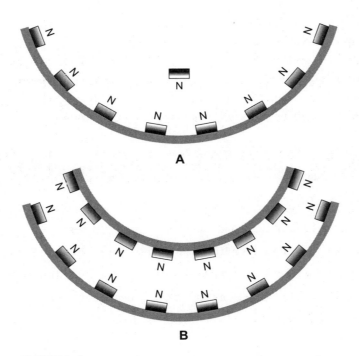

FIGURE 9-1 · When you try to levitate a magnet above a set of other magnets as shown at A, the top magnet flips over and sticks to one of the others. Instability also occurs with two bowl-shaped magnetic structures, one above the other, as shown at B.

Now suppose that you take another 100 magnets and glue them to the outer surface of another mixing bowl, the same shape as but somewhat smaller than, the first bowl, with the north poles facing outward, forming a large magnet with a convex north-pole surface. Suppose you try to set this bowl down inside the first one, as shown in Fig. 9-1B. You hope that the top bowl will hover above the bottom one, but it doesn't. It finds some way to land, off-center, on the bottom bowl. If there are enough magnets on the bowls to prevent a landing inside the bottom bowl, the top bowl will skitter off and land outside the bottom one.

Magnetic levitation cannot be achieved with a set of static (non-moving and non-rotating) permanent magnets. Some instability always exists in such a system, and this instability gets magnified by the slightest disturbance. Samuel Earnshaw proved this fact in the 1800s, and it became known as *Earnshaw's theorem*. Despite the conclusion of this theorem, however, it is possible to obtain magnetic levitation.

? Still Struggling

Earnshaw's theorem derives from a narrow set of assumptions. Engineers can build systems to get around these constraints. Earnshaw's theorem applies only to systems that consist exclusively of permanent magnets with no relative motion among them. In recent years, scientists have come up with dynamic (moving) systems of magnets that can produce levitation. Some such systems are used in railway trains today.

Feedback Systems

Consider the two-bowl scenario of Fig. 9-1B. If you try this experiment, you'll never get the upper bowl to hover indefinitely above the lower bowl in free space. But what will happen if you build a *feedback system* that keeps the upper bowl in alignment with the lower one?

Here's a crude example of how an *electromechanical feedback system* can produce magnetic levitation with the two-bowl scheme. It has an electronic *position sensor* that produces an *error signal*, and a mechanical *position corrector* that operates, based on the error signals from the position sensor, to keep the upper bowl from drifting off center. As long as the upper bowl remains precisely centered over the lower one (Fig. 9-2A), the position sensor produces a zero output

signal. The upper bowl has a tendency to tilt or move sideways because of instability in the system (Fig. 9-2B). As soon as the bowl gets a little off center, the
position sensor produces data that describes the extent and direction of
the displacement. The error data has two components: a *distance error signal* that
gets stronger as the off-center displacement of the upper bowl increases, and a
direction error signal that indicates the direction in which the upper bowl has
drifted. These signals go to a microcomputer, which operates a mechanical

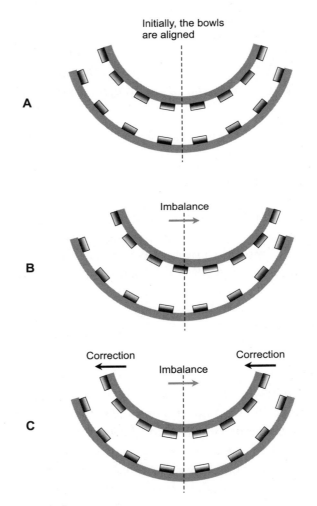

FIGURE 9-2 · A feedback system can keep two magnet-
arrayed bowls properly centered with respect to each
other, producing levitation. At A, the initial aligned situation. At B, an imbalance throws the upper bowl off center.
At C, a correction force brings the upper bowl back to the
center.

device that produces the necessary amount of force, in exactly the right direction, to get the upper bowl back into alignment with the lower one (Fig. 9-2C).

Diamagnetism

Certain substances known as *diamagnetic materials* cause magnetic fields to weaken. A sample of diamagnetic material spreads out, or dilates, the *magnetic lines of flux* when brought into a magnetic field. The phenomenon of *diamagnetism* is the opposite of the more commonly observed effect called *ferromagnetism*. A *ferromagnetic material* such as iron concentrates the magnetic lines of flux when brought into a magnetic field. That effect makes it possible to construct electromagnets.

A diamagnetic substance such as distilled water can, under certain conditions, give rise to magnetic levitation. A diamagnetic object repels either pole of a permanent magnet, just as a ferromagnetic object (such as an iron nail) attracts either pole. This force is too weak to be noticeable with ordinary magnets because they're not strong enough. But levitation can occur if a lightweight diamagnetic object is placed inside a bowl-shaped container arrayed with powerful electromagnets, as shown in Fig. 9-3. A small drop of distilled water, for

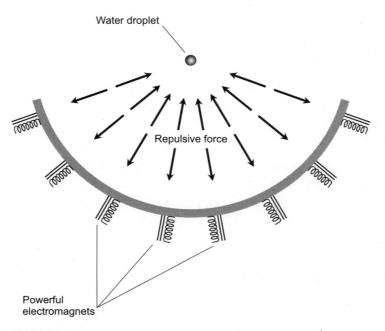

FIGURE 9-3 · Levitation can be produced by the action of powerful electromagnets against a diamagnetic substance such as a droplet of water.

example, can be suspended in midair by this method under controlled laboratory conditions. Unfortunately, this levitation force is far too weak to support heavy objects such as train cars. In order to get a strong repulsive force as a result of diamagnetism, another effect must be exploited: the ability of certain media to become perfect electrical conductors when they get extremely cold.

Superconductors

At temperatures approaching absolute zero (about −273.15°C or −459.67°F), some metals lose all of their electrical resistance. Such an electrical medium is called a *superconductor*, and the phenomenon of zero resistivity (or perfect conductivity) is called *superconductivity*. An electrical current in a superconducting loop of wire can circulate around and around, without growing noticeably weaker, for a long time indeed.

Superconductors make magnetic levitation possible because the magnetic flux is completely expelled from such a medium. The lines of flux are dilated so much that they disappear altogether within a superconductor. Scientists call this phenomenon *perfect diamagnetism*. A more technical name for it is the *Meissner effect*, named after one of its discoverers, Walter Meissner, who first noticed it in the 1930s.

Because of the Meissner effect, superconductors have sparked keen interest among engineers seeking to build vehicles and other machines that take advantage of magnetic levitation. A superconducting diamagnetic medium produces a much stronger repulsive force, for a given magnetic field intensity, than any ordinary diamagnetic sample. The repulsion takes place whether the proximate magnetic pole is north or south, so instabilities associated with simple systems, such as the ones shown in Figs. 9-1 and 9-2, don't occur.

Rotation

We can make an arrangement similar to that shown in Fig. 9-1A work if the upper magnet rotates at a rapid rate, like a hovering top. This rotation makes the upper magnet act as a stabilizing *gyroscope*. A disk made of non-ferromagnetic material (a substance that does not attract magnets) is attached to the upper magnet to increase the gyroscopic effect. Figure 9-4 illustrates this scheme. When the upper magnet spins fast enough, it balances on the mutually repulsive magnetic fields produced by itself and the lower array of magnets.

A system such as the one shown in Fig. 9-4 does not violate Earnshaw's theorem, which applies only to fixed magnets. Nevertheless, the system shown in

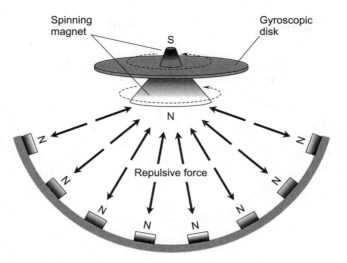

FIGURE 9-4 · A spinning magnet equipped with a gyroscopic disk can levitate above an array of fixed magnets. (The size of the upper magnet is exaggerated for clarity.)

Fig. 9-4 is temperamental. The *angular speed* (rate of rotation) must always remain between certain limits, and the spinning magnet must have just the right shape. After awhile, air resistance will cause the spinning magnet to slow down to the point where the gyroscopic effect fails.

TIP   *A toy called the* **Levitron** *provides a fascinating display of the foregoing principle. You can find an explanation of how it works on the Internet at* **www.levitron.com.**

Oscillating Fields

An object that conducts electric currents but does not concentrate magnetic flux, such as an aluminum disk, exhibits diamagnetism in the presence of an oscillating magnetic field. Such a field can be produced by a set of electromagnets to which high-frequency AC is applied. A suitably shaped, rotating disk made of a non-ferromagnetic metal such as aluminum will levitate above an array of such electromagnets, as shown in Fig. 9-5.

This method of levitation works because of *eddy currents* that appear in the conducting disk. The eddy currents produce a secondary magnetic field that opposes the primary, oscillating field set up by the array of fixed AC

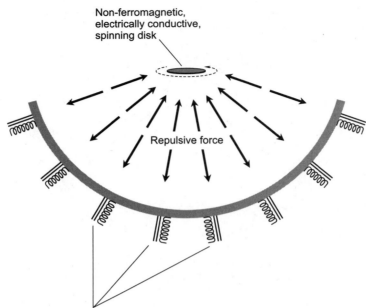

FIGURE 9-5 · A spinning disk, made of non-ferromagnetic material that conducts electric currents, can levitate above a set of AC electromagnets.

electromagnets. In effect, the disk becomes an "antimagnet." The rotation provides a stabilizing, gyroscopic effect, so the disk doesn't flip over. The disk stays centered as long as it is placed along the central axis of the array to begin with, and as long no significant disturbance throws it off center.

? Still Struggling

In the above-mentioned system, the rotation rate of the disk will eventually decay, because of air resistance, to the point that the system becomes unstable. This problem can be eliminated by placing the entire system in a vacuum. (The same thing can be done with the rotating-magnet system described in the previous section.) In theory, this allows the system to operate forever, although in practice it will eventually fail because of inescapable real-world friction and energy loss.

The Maglev Train

High-speed rail transit is one of the most exciting applications of magnetic levitation. Some passenger trains for urban commuters employ this technology, which proponents call *maglev*. The diamagnetic effects of superconductors are most often used for such systems. An adaptation of the rotation scheme, described above, has also been tested on a small scale.

How It Works

In a maglev train, no physical contact exists between the cars and the track. The only friction in the system occurs as a result of the air resistance encountered by the moving cars, which are suspended over a monorail track. A gap of 2 cm to 3 cm (about 1 in) prevails between the train and the track.

The cars in a superconductor maglev train can be supported by either of two geometries shown in Fig. 9-6. In the scheme shown at A, the cars are attached to bearings that wrap around the track. Drawing B shows a system in which the track wraps around bearings attached to the cars. In either arrangement, vertical magnetic fields keep the cars suspended above the track, and horizontal magnetic fields stabilize the cars so they remain centered. Acceleration and braking are provided by a set of *linear motors*, which require an additional set of electromagnets in the track.

An alternative system called *Inductrack* uses permanent magnets in the cars and wire loops in the track. The motion of the cars with respect to the track produces the levitation, in a manner similar to the way a rotating, conducting disk levitates above a set of fixed magnets. This system travels on sets of small wheels as it first gets going. Once it is moving at a few kilometers per hour, the currents in the loops become sufficient to set up magnetic fields that repel the permanent magnets in the train cars. As with the superconductor type maglev train, the Inductrack system uses linear motors to achieve propulsion and braking.

Advantages of the Maglev Train

- Maglev trains are capable of higher speeds than conventional trains.
- Maglev trains make less noise than conventional trains.
- Maglev trains can reduce commute times for people who use trains.
- Maglev trains can make use of low-pollution electrical energy sources.

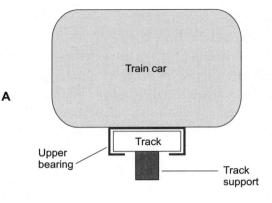

Train motion is perpendicular to page

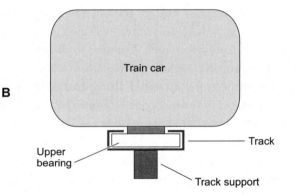

FIGURE 9-6 · Simplified cross-sectional diagrams of maglev train geometries. At A, an upper bearing, attached to the car, wraps around a monorail track. At B, the upper bearing levitates inside a wrap-around track. (In both of these illustrations, the train cars move either straight toward you or straight away from you.)

Limitations of the Maglev Train

- Maglev trains are more expensive than conventional trains.
- Special training is required for maglev maintenance personnel.
- Superconductor maglev trains rely on powerful electromagnets embedded in the track to obtain the levitation. This fact raises the problem of shielding passengers from the strong magnetic fields.

- A sudden power outage will cause the cars in a superconductor maglev train to settle onto the track. This may be dangerous if it occurs at high speed. (In an Inductrack train, the wheels eliminate this danger, allowing the cars to coast to a stop.)

- A high crosswind can disrupt the operation of a maglev train by decentering the cars and causing them to contact the track. Snow or ice on the track can also cause trouble.

PROBLEM 9-1

How can passengers be shielded from the strong magnetic fields in a superconductor type maglev train?

SOLUTION

The train cars, or at least the passenger compartments within them, can be made of a ferromagnetic substance such as steel, which tends to block magnetic lines of flux. Unfortunately, steel has far greater mass per unit volume than aluminum, the other metal commonly used in general construction. Aluminum is not ferromagnetic and would offer no protection against magnetic fields unless it were made to carry high and potentially dangerous electric currents.

PROBLEM 9-2

How can a maglev train negotiate a steep hill or mountain? Won't it fall downhill and settle at the bottom of a valley if no friction exists to provide braking action?

SOLUTION

The linear motors used in maglev train systems can drive the cars up steeper grades than is possible with conventional trains. In addition, the linear motors can provide braking action by switching into reverse, and they can keep the train from falling downgrade by operating against the force of gravity.

The Nuclear-Powered Ship

Nuclear-powered oceangoing vessels have existed for decades. The principal maritime application of nuclear power has historically been confined to submarines and aircraft carriers. However, with the future of the world's oil supply in doubt, *uranium fission* has become a subject of interest for propelling large marine vessels, both military and commercial.

How a Fission-Powered Ship Works

In the type of *nuclear reactor* in use today, elemental uranium gradually decays into lighter elements. The term *nuclear* derives from the fact that the atomic *nucleus* is involved. In fission, the uranium nuclei, consisting of *protons* and *neutrons*, are split apart. Thus, a heavy element is transformed into lighter elements—an ancient alchemist's dream come true!

As fission takes place, energy is released in the form of heat and *ionizing radiation*, also known as *radioactivity*. Some of this radiation consists of high-speed neutrons that break more of the uranium atoms apart, releasing still more energy. If the reaction occurs rapidly enough, an explosion occurs. This is the principle by which the first atom bombs worked. However, the chain reaction can be slowed down and made self-sustaining. When this is done under rigidly controlled conditions, uranium fission can provide large quantities of usable heat energy for long periods of time, and the risk of explosion is essentially nil.

Figure 9-7 is a block diagram of the power plant in a typical nuclear-powered ship or submarine. Heat from the reactor is transferred to a *water boiler* by means of *heat-transfer fluid*, which resembles the *coolant* used in heat pumps, air conditioners, and automobile radiators. This coolant passes from the shell of the boiler back to the reactor through a *coolant pump*. The water in the boiler is converted to steam, which drives a turbine. After passing through the turbine, the steam condenses and is sent back to the boiler by a *feed pump*. The water and heat-transfer fluid are separate, closed systems; neither one comes into physical contact with the other. This separation prevents the accidental discharge of radiation into the environment through the water/steam system.

The turbine can turn the propeller through a *drive system*. The turbine also turns the shaft of a generator that provides electricity for the crew, passengers, and electronic systems. The electricity from this generator can power a motor, or charge a battery that powers a motor, connected to the drive system. The electric motor can provide supplemental or backup propulsion.

Advantages of Fission for Propulsion

- Nuclear fission provides greater "nautical mileage" than any fossil fuel per unit mass, even taking the reactor radiation shield (which is massive) into account.

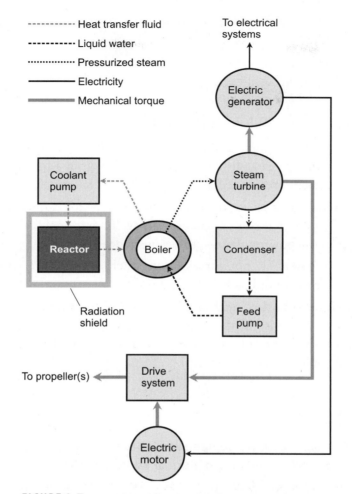

FIGURE 9-7 · Simplified block diagram of a nuclear power and propulsion system suitable for use in large oceangoing vessels.

- Maintenance of nuclear reactors, while critical, need not be done as frequently as refueling and maintenance operations in conventional vessels.
- Nuclear-fission reactors and their associated peripherals can operate in the absence of oxygen. This makes them ideal for use in submarines.
- Nuclear-powered ships can attain higher speeds than conventional submarines.
- When the overall dangers of nuclear-fission propulsion systems are weighed against the proven dangers of conventional systems (oil spills, for example), nuclear fission compares favorably.

- Nuclear-powered ships do not produce greenhouse-gas emissions, CO gas, or particulate pollutants as do fossil-fuel-powered vessels.
- Nuclear energy, in general, can help the world economy wean itself away from reliance on fossil fuels.

Limitations of Fission for Propulsion

- Nuclear fission reactors produce certain waste products that remain radio-active for many years (although most of it decays in a few months). Disposal processes present technical and political challenges.
- Although the risk of accident or sabotage involving a nuclear reactor is small, the potential consequences—leakage of extremely radioactive material into the environment—are serious indeed.
- If certain fissionable nuclear waste products get into the wrong hands, nuclear terrorism or blackmail could result.
- The widespread use of nuclear fission reactors faces opposition from certain groups because of the above mentioned negative factors. This has given rise to public apprehension, particularly in the United States, concerning nuclear energy in general.

PROBLEM 9-3

Why can't nuclear waste be dumped straightaway into the sea? The world's oceans are vast, and the quantity of nuclear waste, even if there are lots of reactors in use, will be small. Wouldn't the oceans dilute the radiation to negligible levels? If that arrangement isn't acceptable, why not put the waste in sealed containers with radiation shielding and drop them to the bottom of the sea, far from land masses or populated islands?

✔ SOLUTION

Nuclear waste has a high radiation output in proportion to its mass. It is difficult to say how low levels of radiation, distributed throughout the world's oceans, might affect marine life and disrupt the food chain from plankton all the way up to humanity. Sealed containers would have to be exceptionally well-made so they would not rupture before the radio-activity had decayed to safe levels.

The Ion Rocket

Hot gases produced by the combustion of flammable fuels, as in conventional rockets, are not the only way to produce thrust for spacecraft. Another way to propel interplanetary and interstellar spacecraft makes use of powerful *linear particle accelerators*. Positively charged atomic nuclei are accelerated to high speed and ejected out the rear of the spaceship, resulting in forward impulse according to the principle of action/reaction.

How It Works

Figure 9-8 is a simplified functional diagram showing how an *ion rocket engine* can work. The ion generator ejects large quantities of ionized gas, such as hydrogen or helium. Positive ions of simple hydrogen consist of individual protons. Positive ions of helium usually contain two protons and two neutrons. (When accelerated to high speed, helium nuclei are sometimes called *alpha particles*.) Any atomic nucleus is, in fact, a positive ion that can be accelerated by negatively charged *anodes*, through which the nuclei pass.

In an ion engine, the anodes carry high voltages, creating powerful electric fields. Successive anodes have higher and higher negative voltages, so they attract positive ions with more and more force. As the ions pass through anode

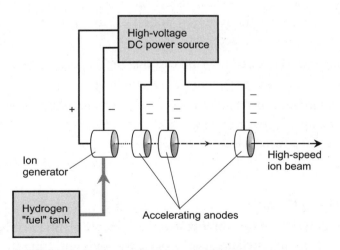

FIGURE 9-8 · Conceptual diagram of an ion rocket engine that produces thrust by accelerating atomic nuclei (ions) to high speed.

after anode, they gain speed, and consequently they gain momentum. When the ions finally emerge from the last anode, they're moving so fast, and they have so much rearward momentum (the product of mass and speed), that the reaction force pushes the spacecraft forward. This force is not great, but it is sustained.

TIP   *Over a long period of time in the near-vacuum of deep space, a spacecraft of this sort can theoretically reach speeds far greater than those attainable by conventional rocket-powered vessels because the fuel lasts far longer.*

Advantages of the Ion Rocket

- Ion engines are efficient. They utilize most of the input energy to produce thrust.
- An ion engine can keep operating for a long time, and thereby can allow a small vessel to achieve high speed, although the rate of acceleration is slow.
- Ion rockets are inherently safe because the fuel does not have to be carried in a form that could combust or explode in outer space.

Limitations of the Ion Rocket

- Linear particle accelerators require large amounts of power in order to function. The only currently existing technology that can provide the necessary power over the required period of time is nuclear fission, in the form of an onboard reactor.
- For long journeys, obtaining and carrying fuel (hydrogen or helium) in the required large amounts could be a problem.
- Because an ion rocket accelerates slowly, it cannot be used as the launch vehicle to put a spacecraft into earth orbit. It is workable only for vessels that are already in space.

PROBLEM 9-4

What factor determines the highest speed attainable by an ion engine operating indefinitely in interstellar space? How is this different from the factor that determines the highest speed attainable by a jet aircraft operating in the earth's atmosphere?

✔ SOLUTION

In a jet-powered aircraft, the maximum attainable speed equals the speed of the ejected matter relative to the engines. Air resistance prevents the craft from moving forward any faster than the exhaust is hurled out rearward. But in space, the situation is different. As long as thrust continues, the speed can increase without limit. The thrust from an ion engine depends on the momentum of the ejected matter. If protons emerge at a constant speed and a constant rate from the rear of a spaceship, the thrust remains constant, and the acceleration, therefore, also remains constant. Given enough time, even modest acceleration, if maintained, can result in extreme speed. The theoretical upper limit is the speed of light according to the *theory of special relativity*. That's approximately 300,000 km/s or 186,000 mi/s.

Fusion Spacecraft Engines

Hydrogen fusion is the nuclear process in which atoms of hydrogen combine, at extremely high temperatures, to form atoms of helium. In this process, energy is liberated. Hydrogen fusion is far more efficient than uranium fission in converting matter to energy. Fusion reactions are also much hotter than fission reactions. This is the type of reaction that takes place deep inside the sun and most other stars.

How They Work

Aerospace engineers have proposed several types of fusion-powered spacecraft as alternatives for obtaining the speeds necessary for long-distance space journeys, using fuel that would have reasonable mass. Well-known designs include the *Orion*, the *Daedalus*, and the *Bussard ramjet*.

In the Orion spaceship, hydrogen fusion bombs would explode at regular intervals to drive the vessel forward. The force of each blast, properly deflected, would accelerate the ship. The blast deflector would have sufficient strength to withstand the violence of the bomb explosions, and it would be made of material that would not melt, vaporize, deform, or erode because of the explosions. The blast deflector would also serve as a radiation shield to protect the astronauts in the living quarters because fusion reactions produce large amounts of deadly X-rays and gamma rays. This idea has been dismissed in recent years because of technical problems, such as the extreme bursts of acceleration (harmful to the crew) and the danger that the bomb blasts could damage or disable the ship.

The Daedalus design would offer a smoother ride, replacing the bombs with a *fusion reactor* that would produce a continuous, controlled "burn." This vessel would use a blast deflector similar to that used in the Orion design. The advantage of the Daedalus ship would be that the acceleration would be steady, rather than intermittent. Daedalus could attain high speed without subjecting the astronauts to bursts of extreme acceleration.

Either the Orion or the Daedalus ships could, according to their proponents, reach 10 percent of the speed of light, or about 18,600 miles per second (30,000 kilometers per second). This speed would make it possible to reach the nearest star outside of our own Solar System, Proxima Centauri, and return within a few human life spans. The ships would accelerate for the first half of the journey, and decelerate for the second half. With the Daedalus design, deceleration could be obtained by turning the vessel around so that its tail points forward, causing the exhaust to produce a rearward impulse.

The Bussard ramjet is the most intriguing nuclear fusion design of the three mentioned here. It resembles the Daedalus, but it would not have to carry nearly as much fuel. Once the ship got up to a certain speed, a huge "scoop" in the front could (hopefully) gather up enough hydrogen atoms from interstellar space to provide the necessary fuel for sustained fusion reactions. Figure 9-9 is a simplified diagram of this type of spaceship. The greater the speed attained,

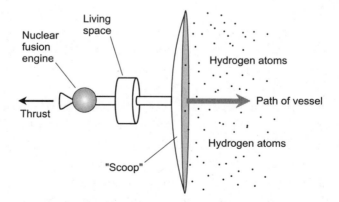

FIGURE 9-9 · The Bussard ramjet gets its fuel by scooping up hydrogen atoms from space and using them to fuel a fusion engine.

the more hydrogen the ship could sweep up, thus helping it go even faster. The ship would decelerate by directing the exhaust out of a forward-facing nozzle in the center of the scoop.

The Bussard ramjet would work especially well in gaseous nebulae, provided those regions were not too peppered with meteoroids and asteroids. Some scientists think that this type of spaceship could attain speeds great enough to take advantage of a phenomenon called *relativistic time dilation*. The astronauts on board the ship would age more slowly than they would if the vessel were traveling at conventional speeds, and this effect could increase the distance reachable within a human lifetime.

TIP *Time dilation becomes significant at speeds greater than 90 percent of the speed of light: roughly 270,000 km/s or 170,000 mi/s. As a spacecraft approaches the speed of light, the time-dilation factor increases without limit. This phenomenon has proven fertile territory for science-fiction authors!*

Advantages of Fusion Spacecraft Engines

- Hydrogen is the most abundant element in the universe. It exists in free form in outer space. It could provide an unlimited supply of fuel, or at least minimize the amount that would have to be carried on board a space vessel.

- Hydrogen fusion produces a tremendous amount of energy from a small amount of matter.

- Hydrogen fusion produces essentially no radioactive waste. The only significant by-product is helium. No CO, CO_2, SO_x, NO_x, or particulate pollutants are generated either.

- A fusion reactor could provide all the energy necessary to operate the electrical, electromechanical, and electronic systems on board the vessel.

Limitations of Fusion Spacecraft Engines

- The crew would have to be protected from the radiation produced by the fusion reaction. This would necessitate massive shielding, reducing the attainable acceleration for a given amount of thrust.

- Even though a fusion engine may allow a vessel to attain high speed, it is unrealistic to suppose that humans will ever use such spaceships to "roam the galaxy" as they do in science-fiction movies and television shows. Some entirely different means of propulsion, as yet unknown, will have to emerge if we ever want to see that sort of thing.

- The materials in the blast deflector would have to withstand extreme temperatures, as well as immense mechanical stresses, for a prolonged period of time.

- If near-light speed were attained to take advantage of time dilation, the earth would age more rapidly than would the vessel's crew. If the astronauts returned to the "home world" after a long journey of this sort, they would find themselves in the distant future—permanently.

PROBLEM 9 - 5

Could a hydrogen-fusion space vessel be launched directly from the earth's surface? If so, wouldn't the blast from the engines cause destruction and deadly radiation near the launch site?

SOLUTION

Most proposals for hydrogen-fusion-powered spacecraft envision a conventional rocket launch vehicle to place the ship in an earth orbit. The fusion engines would be activated at an altitude of several thousand kilometers. Therefore, no danger would exist for people on the surface.

The Solar Sail

The sun emits some of its energy in the form of high-speed subatomic particles. These particles create a so-called *solar wind* that rushes outward through our Solar System. Maybe we can take advantage of this solar wind to move a spaceship along, in much the same way as moving air molecules on earth propel sailboats!

How It Works

A spaceship with a *solar sail* would require an enormous sheet of reflective fabric or foil. This sheet would be attached to the living quarters (Fig. 9-10). The range of travel would be limited to our Solar System.

With a solar sail, it would be easier to move away from the sun than towards it, for obvious reasons. But, just as sailing ocean vessels can travel into the wind by *tacking* (taking a zigzag path), "solar sailors" would be able to navigate in any direction, given sufficient solar wind speed. To move in closer to the sun, they would follow a spiraling path inward, a sort of "three-dimensional tack," the

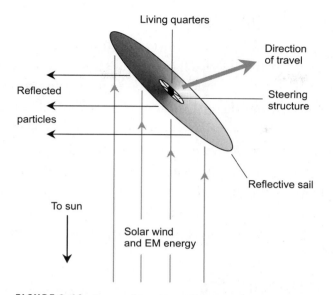

FIGURE 9-10 · A spaceship with a solar sail rides on the solar wind, just as a sailboat rides on atmospheric wind.

direction of travel subtending an angle of slightly less than 90° with respect to the solar wind.

The solar sail requires no on-board fuel, at least in the ideal case. However, navigation would be tricky. Sudden *solar flares* would produce a dramatic increase in the solar wind because of the large number of relatively massive particles ejected during such an event. Near any planet with a magnetic field, including the earth, the particles are deflected and the solar wind does not necessarily "blow" away from the star.

? Still Struggling

The classic ocean sailing ships did not come in all the way to the beach, but dropped anchor in deep water and sent small boats to shore. During space voyages, the main ship could draw in its solar sail and "drop anchor" by falling into a planetary orbit. Landings could be made with small shuttles that would be sent out from, and return to, the main ship.

Advantages of the Solar Sail

- The solar sail is a passive device. It does not require the ship to carry, or collect, any fuel. (However, a backup propulsion system, such as a set of conventional rockets or a nuclear-fusion engine, would be a good idea.)
- The mechanical and thermal stresses on a solar sail are far smaller than those in a conventional rocket engine or a nuclear-fusion engine.
- A solar sail produces no radiation or waste products.
- Once deployed, a solar sail would require no maintenance except for occasional repairs of punctures produced by meteoroids.
- A solar sail could be partially coated with thin-film photovoltaics (solar panels) to provide electrical energy for onboard hardware and life-support systems.

Limitations of the Solar Sail

- The solar sail relies on the stream of particles from the sun. It might not work in interstellar space for lack of a dominant star to produce a defined and predictable particle wind.
- The maximum speed attainable with a solar sail would be much lower than the maximum speed of a spaceship powered by an ion or fusion engine.
- A solar sail, because of its large size, would be difficult and awkward to deploy.
- Sudden solar flares and planetary magnetic fields would complicate the navigation of a ship using a solar sail. It would be something like sailing a marine vessel through a region with shifting winds and ocean currents.

PROBLEM 9-6

Wouldn't the huge surface area of a solar sail create friction with the rarefied gases normally present in interplanetary space, rendering the system impracticable?

✔ SOLUTION

There would be some friction between the solar sail and atoms of matter—mainly hydrogen and other gases—in space. However, the force produced by the solar wind should be far greater than the resistance offered by this friction, at least in the inner solar system. But of course, we'll never know how well a solar sail will work until it is tested "in the field"!

QUIZ

Refer to the text in this chapter if necessary. A good score is eight correct. You'll find the correct answers listed in the back of the book.

1. **In a nuclear fission reaction,**
    A. lighter elements combine to form heavier ones.
    B. heavier elements get split into lighter ones.
    C. electrons and protons merge to form neutrons.
    D. compounds break apart into their individual atoms.

2. **Solar wind comprises**
    A. ultraviolet radiation.
    B. air molecules.
    C. subatomic particles.
    D. visible light.

3. **Which of the following factors constitutes an advantage of hydrogen fusion over nuclear fission?**
    A. Fusion produces essentially no atomic waste.
    B. Fusion reactions don't get as hot as fission reactions do.
    C. Fusion reactions don't produce any dangerous radiation.
    D. All of the above

4. **In a nuclear-powered submarine, the turbine that actually turns the propellers is directly driven by**
    A. electrons.
    B. high-speed protons.
    C. steam.
    D. X rays.

5. **The Daedalus spacecraft would likely attain maximum speeds of about**
    A. half the speed of light.
    B. 18,600 miles per hour.
    C. 30,000 kilometers per second.
    D. 186,000 kilometers per hour.

6. **Earnshaw's theorem tells us that in order for a set of magnets to produce levitation, at least some of the magnets must be**
    A. unipolar.
    B. superconductors.
    C. powered by electricity.
    D. in motion.

7. Which type of material concentrates magnetic lines of flux the most?
   A. Ferromagnetic
   B. Diamagnetic
   C. Hypermagnetic
   D. Quasimagnetic

8. Which type of material actually weakens magnetic fields by dilating the lines of flux?
   A. Ferromagnetic
   B. Diamagnetic
   C. Hypermagnetic
   D. Quasimagnetic

9. Perfect diamagnetism is also known as
   A. Meissner effect.
   B. Earnshaw effect.
   C. levitation effect.
   D. superconductivity.

10. An ion rocket engine's exhaust, which produces the impulse according to the action/reaction principle, usually comprises
    A. extremely hot gases.
    B. electrons.
    C. protons or helium nuclei.
    D. neutrons.

chapter 10

Electricity from Fossil Fuels

As nonrenewable sources of electrical energy become less abundant and more costly, the nations of the world will have to exploit alternatives. Right now, however, a significant proportion of the world's electricity comes from generators driven by turbines that ultimately get their energy from fossil-fuel combustion.

CHAPTER OBJECTIVES

In this chapter, you will

- Find out how coal power plants work, and why they still exist.
- Discover why long-distance power transmission lines work best when they carry high voltages.
- Compare the assets and limitations of coal, oil, and methane-fired power plants.
- Learn how small fossil-fuel-powered generators can provide emergency backup power when the utilities fail.

Coal-Fired Power Plants

Although coal is a nonrenewable resource, plenty of it still resides in the ground. In the first years of the twenty-first century, coal experienced a resurgence in usage, as supplies of oil and natural gas, the other combustibles most often used to generate electricity, grew short. Coal is not a permanent solution to human-kind's energy problems, but it's available for our responsible use as we work toward developing more lasting alternatives.

How They Work

In recent years, coal has been branded as a dirty, inefficient energy source, the mining of which destroys the landscape, and the burning of which pollutes the atmosphere. However, a modern coal-burning electric generating facility is a sophisticated operation and can run efficiently without ruining the environment. Coal is mined in locations that are usually some distance away from generating plants, so the coal must be transported, in most cases by rail. The coal is cleaned and *degassed* at the mining site prior to being loaded onto *coal trains* that trundle endlessly across the countryside.

Figure 10-1 is a simplified functional diagram of a coal-fired electric generating plant. Once the coal arrives at the generating site, it is loaded into a large *hopper* (although excess may pile up into little black hills at the site). From the hopper, the coal moves into a *pulverizer* that grinds it into dust. A *blower* (not shown) forces air laden with coal dust into the *combustion chamber*, where the coal dust burns. The resulting heat turns liquid water into pressurized steam in the *boiler*. The water supply is purified to get rid of minerals that would otherwise build up in the system over time. The steam drives a *steam turbine*, which turns the shaft of one or more *generators*. The steam from the turbine is cooled and condensed back into liquid water by a *condenser*. A *feed pump* returns the liquid water to the boiler.

TIP   *The electricity produced by an electric generator manifests itself as AC, standardized at a frequency of 60 hertz (Hz) in the United States. In some countries the frequency is 50 Hz. One hertz is equivalent to one complete wave cycle per second. The voltage of this AC is increased to several hundred thousand volts by a* huge *step-up transformer* connected to each generator. The high-voltage electricity goes to the *transmission lines* for distribution. Large coal-fired power*

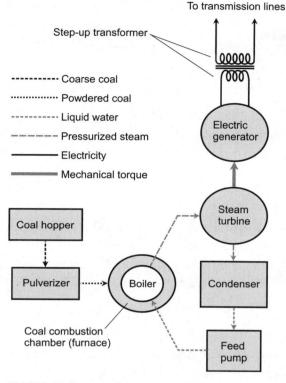

FIGURE 10-1 · Simplified functional block diagram of a coal-fired electric generating plant using a boiler and steam turbine.

plants can supply many megawatts (MW) of electrical power, enough to run thousands of households and businesses.

Why Such High Voltage?

Have you ever wondered why long-distance power transmission takes place at high voltages, necessitating massive towers and gigantic insulators? Why, you might ask, can't electricity be transmitted at a low voltage in heavy-duty wires running between modest structures, or even underground?

For any specific, fixed amount of power ultimately consumed by the *load* (the combination of all the end users), the current in an electrical transmission line goes down as the voltage goes up. Reducing the current (by increasing the voltage) lowers the amount of *power loss* that occurs in the line's conducting wires. Recall from your basic electricity course the following formula:

$$P = EI$$

where P represents the power in watts, E represents the voltage in volts, and I represents the current in amperes. By rearranging this formula, we can see that at any given power level, the current varies inversely with respect to the voltage:

$$I = P/E$$

The power loss in a transmission line (which always results from the wire resistance) is proportional to the *square* of the current. This loss constitutes power that never reaches the end users; it heats up the transmission line's wires and, in effect, goes completely to waste. The following relationship holds:

$$P = I^2 R$$

where P represents the power in watts, I represents the current in amperes, and R represents the resistance of the wire in ohms.

Engineers can't do much about the wire resistance or the power consumed by the load in a power transmission system, but they can maximize the voltage of the electricity that they feed into the line. In this way, engineers can minimize the current that a transmission line must carry in order to provide for the needs of the end users.

Imagine that we step up the voltage into a transmission line by a factor of 10, while the combined loads at the end of the line draw constant power. The increase in voltage reduces the current to 1/10 of its previous value. As a result, the power loss goes down to $(1/10)^2$, or 1/100, of its former level—in the very same transmission line.

TIP *The use of a step-up transformer in a single location costs a lot less than the alternative: stringing up wires with 100 times as much cross-sectional area as the existing transmission line, thereby consuming 100 times as much metal and weighing 100 times as much, necessitating the use of vastly larger, more expensive, more unsightly, and more dangerous supporting towers.*

? Still Struggling

Does it scare you to contemplate a power transmission line that carries half a million volts? If so, consider this: The health risk from power lines (the extent of which has given rise to debate) comes from the *magnetic fields* they generate, not from the voltages they carry. The strength of the magnetic fields varies in proportion to the current, not the voltage. If that big transmission line through the outskirts of your town operated at 500 volts rather than 500,000 volts, the magnetic field near it would have far greater intensity, and the potential for harmful effects would be correspondingly greater.

Along the Line

Extreme voltage is good for *high-tension* (meaning high-voltage) power transmission, but the average consumer can't make any practical use of it. The wiring in a high-tension system must be done using precautions to prevent *arcing* (sparking) and short circuits. Personnel must be kept at least several meters away from the wires. Electrocution by a typical high-tension power line would mean instant death.

Can you imagine trying to use an appliance, say a computer or a television set, directly with an electrical system that delivers a couple of hundred thousand volts? You'd get killed before you could even put the plug into the outlet. At such a voltage, the electrons would jump through the air to your hand, pass through your body to whatever electrical ground they could find, and flash-fry you from the inside out. You might as well get hit by a lightning bolt.

At various points near groups of end users, medium-voltage power lines branch out from the major lines, and *step-down transformers* are used at the branch points. The step-down transformers reduce the voltage in the power lines. The lower-voltage lines fan out to more step-down transformers, leading to even lower-voltage lines. Finally, several lines from each transformer serve individual buildings. Each transformer must contain wires thick enough to withstand the maximum electrical current demanded by all the consumers that it serves.

Sometimes, such as during a heat wave, the demand for electricity rises above the normal peak level. This demand "loads down" the circuit to the point that the voltage drops several percent. Then we experience a so-called *brownout*. If consumption rises further still, a dangerous current load is placed on one or more intermediate power transformers. Circuit breakers in the transformers protect them from destruction by opening the circuit, and we have to contend with a *blackout*.

At individual homes and buildings, transformers step the voltage down to either 234 V or 117 V. Usually, 234-V electricity is provided in the form of three sine waves, called *phases*, with each phase appearing at one of the three slots in the outlet (Fig. 10-2A). This voltage is commonly employed with heavy appliances, such as electric ovens and stoves, electric furnaces, and electric laundry washers and dryers. A 117-V outlet supplies single-phase AC voltage between two of the three slots in the outlet. The third opening leads to an *earth ground* (Fig. 10-2B).

Advantages of Coal for Electrification

- Coal is abundant, and plenty of it exists in the United States. This abundance is of obvious benefit to the American economy in the face of uncertain supplies and unpredictable prices for oil and methane.
- Modern coal-fired power plants are efficient, and produce much less pollution than their "olden-days" counterparts did.
- The furnaces in power plants that use pulverized coal are flexible. They can burn all grades of coal, from lignite (soft coal) to anthracite (hard coal), and they also allow for the combustion of oil and/or methane.

Limitations of Coal for Electrification

- The supply of usable coal, while vast, is not infinite. At best, it can provide temporary relief from, but not a permanent solution to, the world's long-term electrification problems.

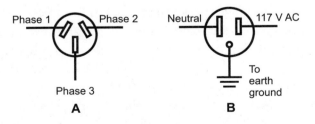

FIGURE 10-2 · At A, an outlet for three-phase AC. At B, a conventional single-phase AC utility outlet. (These symbols apply in the United States. International symbols may vary.)

- The combustion of coal, while cleaner than it used to be, generates CO_2 (a known greenhouse gas), CO, sulfur oxides (SO_x), nitrogen oxides (NO_x), and mercury compounds. Emission-control devices mitigate the air pollution when properly employed, but in countries with emerging economies, these devices are not always used.

- Coal mining leaves long-lasting scars on the landscape, and can result in runoff of toxic substances, such as lead, mercury, and arsenic.

- Frequent coal trains impede road traffic in cities without enough railway overpasses and underpasses. This problem presents itself not only as a nuisance, but a potential danger if it delays the passage of emergency vehicles, such as ambulances, police cars, and fire trucks.

- Frequent long, heavy trains produce noise that can annoy people living near the tracks, and can also reduce the values of properties near the tracks.

- Rail transport requires fuel, increasing the overall pollution involved with coal-generated electricity, and decreasing the overall efficiency by requiring that one form of fuel (such as diesel) be consumed to get the other (coal) to the points of use.

- The water used in the boiler of a coal-fired power plant accumulates pollutants. When this water is replaced, the pollutants must be safely disposed of, increasing the cost of operation.

▢ PROBLEM 10-1

Why can't huge coal-fired power plants be constructed where the mines are located, doing away with the need for coal trains?

✔ SOLUTION

In theory, this could be done. However, much of the available coal is located far from major population centers. This would require extremely long transmission lines. The cost of constructing, operating, and maintaining these transmission lines would exceed the cost of running coal trains to smaller power plants located closer to the end users.

Oil-Fired Power Plants

In the United States, oil is employed mainly for heating and propulsion, and is not widely used to generate electricity. In some respects, oil resembles "liquid coal." Oil is pumped out of the ground rather than dug or scraped out, but it must be transported from the places where it is obtained to the places where it is refined, and thence to the places where it is burned.

How They Work

Three configurations are used in oil-fired power plants: the *conventional steam system*, the *combustion-turbine system*, and the *combined-cycle system*. In all instances, oil is transported to the generating plant from refineries, and is usually stored in tanks on-site. Transportation from the refineries to the power plant can be done in large waterborne oil tankers, by means of trains or trucks, or by means of pipelines.

In a conventional steam oil-fired power plant, the fuel is burned in much the same way as is done in an oil-fired home heating furnace, but on a larger scale. The heat from the combustion boils water. The resulting hot water vapor drives a steam turbine (Fig. 10-3). Except for the physical nature of the fuel, this system is similar to a coal-fired power plant.

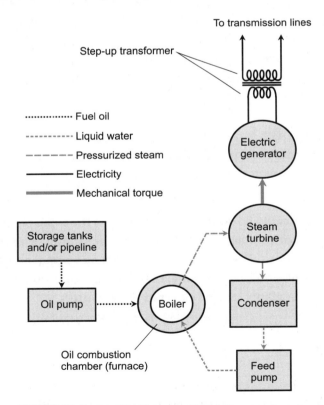

FIGURE 10-3 · Simplified functional block diagram of an oil-fired electric generating plant using a boiler and steam turbine.

In a combustion-turbine, oil-fired power plant, the burning of the fuel produces fast-moving exhaust that passes through a *gas turbine*, which resembles a windmill that one might design for operation in hurricanes or tornadoes (if one were predisposed to do such a thing). The turbine drives the shaft of an electric generator as shown in Fig. 10-4.

A combined-cycle, oil-fired power plant consists of the same components as a combustion-turbine system, but in addition, the hot exhaust is used to produce steam in a water boiler, and this high-pressure steam drives a second turbine. In this way, more of the energy is recovered, and the efficiency of the system is increased.

Advantages of Oil for Electrification

- Oil is a relatively safe fuel. An oil leak or spill can cause a fire, but it does not pose the danger of an explosion as flammable gases do.

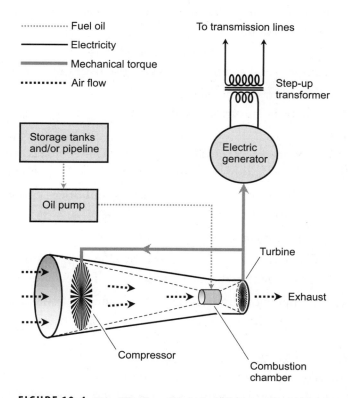

FIGURE 10-4 · Simplified functional block diagram of an oil-fired electric generating plant using a combustion turbine.

- Oil is a high-density fuel. A moderate-sized on-site oil tank can hold enough fuel to produce a large amount of electrical energy.

- Fuel oil can be mixed with biofuel. Most oil-fired power plants can be designed to efficiently burn such a mixture, known as *hybrid fuel*.

- Steam-turbine, oil-fired power plants can be modified to burn coal and/or methane if necessary.

- Modern oil-fired power plants produce less pollution than they did in the "olden days," largely because of the introduction and use of emission-control hardware in the exhaust systems.

Limitations of Oil for Electrification

- Fuel-oil combustion, although cleaner than it once was, produces air and water pollutants similar to those produced by the combustion of coal. Emission-control systems can help reduce this pollution, but only if they are kept in proper working order. In some emerging countries, emission control is not affordable, and the result is increased pollution and green-house-gas production.

- The price of fuel oil is directly related to the price of crude oil. This price can spike rapidly and can be expected to rise over the long term.

- Much of the world's crude oil comes from politically unstable parts of the world, so there is an ongoing risk of sudden and unpredictable supply disruptions.

- Temporary reductions in the oil supply for the United States can result from natural events, such as hurricanes, earthquakes, and pipeline corrosion.

- Speculation about oil production, transport, and prices can artificially inflate the price from time to time.

- Oil leaks and spills can harm the environment.

- The transport of crude and refined oil by ship, rail, and truck consumes energy. This, in effect, reduces the efficiency of the whole process.

- The world's supply of crude oil is finite and nonrenewable. Sooner or later, we will run out of it!

PROBLEM 10-2

Do combustion type power plants require cooling systems to keep the components from overheating? Does the heat discharged from such a power plant have environmental effects?

✔SOLUTION

The answer to both questions is "Yes." In the interest of simplicity, cooling systems are not shown in the block diagrams here. Combustion type power plants are located near bodies of water in order to provide a plentiful source of "coolant." (Ocean water must be desalinated for use in cooling systems.) The warm water discharged from the cooling system eventually returns to its source, raising the temperature of the lake, river, or ocean. This affects aquatic and marine life in the vicinity of the power plant, but this is not necessarily a bad thing. For example, if a power plant next to a northern river keeps the river open in winter, wildlife may flock to the area during the cold season.

Methane-Fired Power Plants

Methane is a component of natural gas, a nonrenewable fossil fuel. Underground reservoirs of natural gas are plentiful but finite. In the late twentieth century, methane began to replace coal, oil, and nuclear fission on a large scale for the purpose of electric power generation in the United States. Reliance on methane will likely increase during the first part of the twenty-first century. But supply and price problems have been experienced with methane, just as has been the case with oil.

Recovery and Transportation

Natural gas is recovered from underground reserves and refined to obtain methane. Another way to get methane is to extract it from coal beds. Because North America has vast coal reserves, this technology can help increase the supply of methane in the medium term. In the United States, proposals to exploit coal beds for methane production encounter opposition from environmental groups and some local residents. Other residents welcome such development because of the potential benefits to their local and regional economies.

There's still another way to get methane: from the decomposition of certain biological substances such as animal waste. Methane from *biomass* contributes relatively little to the overall supply at the time of this writing, but it's a renewable resource.

The methane, once extracted and refined, can be transported to generating plants by pipelines. Alternatively, the gas can be liquefied for transportation by rail and truck, and for on-site storage in tanks. The transport of methane in any form is more dangerous than the transport of coal or oil. Coal transport can be a nuisance and an oil spill can contaminate soil and drinking water, but a methane leak can cause a deadly explosion and flash fire.

The Combined-Cycle System

Coal-fired, steam-boiler power plants can be modified to burn methane. However, combustion-turbine and combined-cycle electric generating systems are more commonly used with methane. A well-designed combined-cycle methane system can operate at higher efficiency than older systems using boilers and steam turbines.

Figure 10-5 is a simplified block diagram of a combined-cycle, electric-generating plant that burns methane. The fuel is supplied to a combustion chamber in a machine that resembles a gigantic jet aircraft engine. A *compressor* drives air into the system. The methane combustion heats this air, causing high-speed

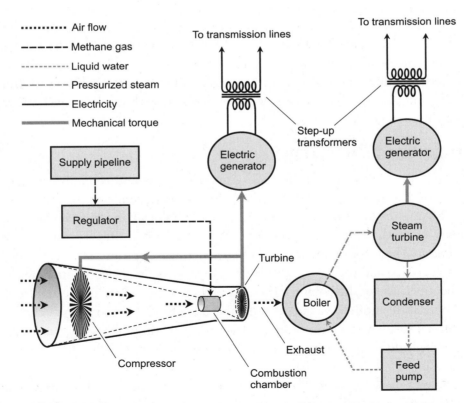

FIGURE 10-5 · Simplified functional block diagram of a combined-cycle methane-fired electric generating plant.

exhaust to pass through a gas turbine. The gas turbine provides power to keep the compressor running, and also drives the shaft of an electric generator. The exhaust supplies heat energy to convert water into steam in the boiler. The steam passes through a steam turbine that drives a second electric generator.

By taking advantage of the energy contained in the exhaust from the combustion turbine, rather than simply letting it escape into the atmosphere, efficiency is optimized and *heat pollution* is reduced. The process resembles (but is a more sophisticated version of) energy reuse by the exhaust converters found in modern wood stoves.

TIP   *The energy reuse scheme described above isn't perfect. The steam from the boiler must be recondensed, releasing heat energy into the environment. In addition, the steam-turbine system requires a source of water, necessitating a periodic discharge of waste water when the cooling apparatus is flushed and its water supply is replaced.*

Advantages of Methane for Electrification

- Methane combustion produces relatively low particulate, NO_x, SO_x, and CO emissions.

- Methane can serve as a medium-term transition fuel as electric generating plants evolve to take increasing advantage of alternative resources, such as wind, solar, hydroelectric, tidal, and geothermal energy.

- Methane-fired, combined-cycle power plants are efficient.

- Methane is readily available in most cities and towns, near the end users of the electricity. This means that methane-fired power plants are easier to site than coal- or oil-fired plants, and there is less need for long high-voltage electric transmission lines.

- An uninterrupted supply of methane can be provided by underground pipelines, reducing the need for energy-consuming trains and trucks to transport the fuel from the refinery to the electric-generating plant.

- Methane-fired power plants might be modified to burn hydrogen gas if that fuel source becomes available in quantity at a reasonable price.

Limitations of Methane for Electrification

- When methane leaks into the air, the mixture of gases becomes explosive. In most locales, methane gas is given an artificial scent that is easy to recognize. The odor can alert people to the existence of gas leaks.

- In recent years, price "spikes" and supply disruptions have marred the reputation of methane as a reliable source of energy.
- The world's supply of naturally occurring methane is finite, and it is not renewable. (However, methane derived from biomass is renewable.)
- Exploration for, and recovery of, natural gas can adversely impact the environment by causing erosion, accelerating runoff, and increasing the risk of mudslides and floods.
- If not responsibly done, exploration for, and recovery of, natural gas may disrupt wildlife habitats and migration routes.
- The combustion of methane produces CO_2, a known greenhouse gas. Methane itself is a greenhouse gas too. Any methane that leaks out in exploration, recovery, and production contributes to the overall problem.

? Still Struggling

In recent years, the use of *induced hydraulic fracturing* (or "fracking"), in which methane (and to some extent oil) is forced from the earth by pumping water deep into the ground, has caused controversy because of fears that it might pollute drinking water and induce instability in the earth's crust, increasing the risk of localized earthquakes. As people's demand for methane increases, the energy industry will face pressure to develop new ways to get that gas out of the ground. We should not be surprised if time reveals adverse results from this process. As with any energy-production and consumption process, we have to weigh its benefits against its hazards.

PROBLEM 10-3

Methane is widely used for home heating. If electric power plants rely increasingly on methane, won't it put a strain on the supply, causing ever-more-severe problems with wintertime price volatility and spot shortages?

✔ SOLUTION

This is an important issue, and it is one of the main arguments used by advocates for increased construction and deployment of nuclear, hydroelectric, solar, wind, geothermal, and even coal-fired power generating facilities.

On-Site Combustion Generators

Small and medium-sized *combustion generators* are available for use in homes, businesses, hospitals, and government buildings. Some generators are also suitable for use by campers. For people living in remote areas, a combustion generator may constitute the primary, if not the only, source of electricity.

How They Work

Figure 10-6 is a functional diagram of a typical combustion generator. The unit shown in this example provides 117 V AC, which is the usual output of portable generators used to power small appliances such as lamps and radios. Larger generators also supply 234 V AC for heavy appliances such as electric ranges and laundry machines. The engine can range in size from a few horsepower

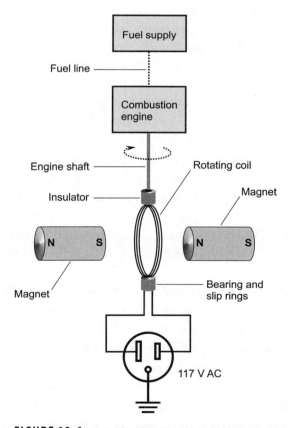

FIGURE 10-6 · Simplified functional block diagram of a small or medium-sized onsite combustion generator.

(comparable to the engine in a lawn mower) to hundreds of horsepower (comparable to the engines in trucks, tractors, and construction equipment). Most small generator engines burn gasoline. Larger ones use diesel fuel, propane, or methane.

In the AC generator, a coil of wire, attached to the shaft of the combustion engine, rotates inside a pair of powerful magnets. If a load is connected to this coil, an AC voltage appears across its end terminals as each point in the wire coil moves across *lines of flux* produced by the magnets, first in one direction and then in the other direction, over and over. (In an alternative arrangement, the magnetic poles revolve around the wire coil, which remains fixed). The AC voltage that a generator can produce depends on the strength of the magnets, the number of turns in the wire coil, and the speed of rotation. The AC frequency depends only on the speed of rotation. In the United States, this speed is 3600 revolutions per minute (3600 rpm) or 60 revolutions per second (60 rps), so the output frequency is 60 cycles per second (60 Hz). In order to maintain a constant rotational speed for the generator under conditions of variable engine speed, mechanical regulating devices are used.

When a load is connected to the output of a generator, it becomes mechanically harder to turn the generator shaft, as compared with the situation when nothing is connected to the output. As the amount of electrical power demanded from a generator increases, so does the mechanical power required to drive it, and therefore, the amount of fuel consumed per unit of time. The electrical power that comes out of a generator is always less than the mechanical power required to drive it. The lost energy manifests itself as heat in the generator components. The *efficiency* of a generator is the ratio of the electrical power output to the mechanical driving power, both measured in the same units such as watts (W) or kilowatts (kW), multiplied by 100 to get a percentage. No generator is 100-percent efficient, but a good one can come fairly close.

TIP *A typical small gasoline-powered generator provides from 1 kW to 5 kW of electrical power. Such a generator, if not well designed, can cause problems when you try to run sensitive electronic equipment from it. However, a well-engineered generator, even the small gasoline-burning type, will work fine with computers and other sophisticated systems, as long as you keep it in good working order.*

TIP *Medium-sized diesel-, propane-, and methane-fueled generators can supply several tens of kilowatts, and can power an entire home, business, or agency. Large institutions typically have multiple generators. These machines, if properly* *operated and maintained, can operate all kinds of equipment, even sensitive and complex medical devices.*

Still Struggling

Some small-scale generators for residential use circumvent the need for constant motor speed by converting the generated AC to regulated DC, and then using a power inverter to generate clean 60-Hz AC from that DC. If the motor speeds up (as it will under reduced load) or slows down (as it will under increased load), the DC voltage does not change because the regulator holds it constant. Therefore, the AC output frequency stays at a constant 60 Hz.

WARNING! *An on-site standby generator must be run only when the wiring in the building has been completely separated from the electric utility by means of a* double-pole, double-throw *(DPDT)* isolation switch. *Otherwise,* backfeed *can occur, in which the electricity from the generator gets into the utility lines near the home or business where the generator operates. Backfeed can endanger utility workers and damage electrical system components. In the United States. the DPDT isolation arrangement is required by the* National Electric Code *(NEC).*

Advantages of On-Site Combustion Generators

- A well-maintained and properly operated on-site generator can eliminate much, if not all, of the inconvenience associated with utility power blackouts.

- In a critical setting such as a hospital, on-site standby generators can save lives.

- When properly installed, operated, and maintained, a combustion generator can provide power any time it is needed, and for as long as it is needed. This does not always hold true for alternative electric energy sources, such as stand-alone photovoltaic (solar) systems or wind-driven systems.

- Well-designed and well-maintained combustion generators are rugged, reliable, and reasonably efficient.

- Combustion generator technology is adaptable. For example, generators fueled by petroleum diesel can be adapted to burn biodiesel. Methane-fueled generators can burn methane derived from a biological process as well as the more familiar product of natural-gas or coal refining.

Limitations of On-Site Combustion Generators

- Combustion generators produce exhaust with the same pollutants that come from combustion type furnaces and motor vehicles.
- On-site combustion generators can be dangerous. All instructions and electrical codes must be strictly followed.
- An on-site combustion generator must be provided with an uninterrupted supply of fuel for as long as its service is needed. In the case of gasoline, diesel fuel, and propane, this requires on-site storage tanks.
- Methane-powered generators will not work if the gas utility is interrupted. Methane utilities routinely shut off the gas during violent disasters to prevent flash fires and explosions in the event a line is ruptured.

PROBLEM 10 - 4

I'd like to build an energy-independent "get away from it all" house in the Red Desert of Wyoming. It's cold there in the winter, but the sun shines a lot, and the wind is a reliable resource as well. I'd like to use an on-site combustion generator to supplement a stand-alone solar-power system and a stand-alone wind-power system. Is that a workable idea?

✔SOLUTION

With all three of these power sources available, you should never have to worry about an electricity blackout! Be sure there's plenty of fuel available for the combustion generator in case there is a prolonged spell of cloudy, windless weather. Keep in mind that a comprehensive system of this sort, while offering independence from the electric utility, will cost you a lot of money up front. In addition, fuel and maintenance will impose ongoing expenses. You should size each system to ensure that you will always have the amount of electrical power you need. The services of a competent engineer will be required if the systems share any wiring. Otherwise, you might have trouble with electrical conflicts (known as *bucking*) between different systems. This condition, which can damage system components as well as equipment connected to the electrical wiring, occurs when the waves from two or more AC sources are not kept in precise phase coincidence (lock-step) with each other.

QUIZ

Refer to the text in this chapter if necessary. A good score is eight correct. You'll find the correct answers listed in the back of the book.

1. **High voltage offers an advantage for long-distance power transmission because it**
   A. maximizes the magnetic field strength.
   B. reduces the danger of electric shock.
   C. reduces the amount of energy lost as heat in the wire resistance.
   D. All of the above

2. **Assuming constant power demand from the end users and constant power-line resistance, doubling the voltage in a long-distance transmission line would**
   A. double the power loss in the line.
   B. quadruple the power loss in the line.
   C. cut the line loss to half its former amount.
   D. cut the line loss to 1/4 of its former amount.

3. **Methane-fired power plants might someday be made more efficient, and even less polluting than they already are, by modifying them to burn**
   A. gasoline.
   B. hydrogen.
   C. wood.
   D. coal.

4. **The prospect of frequent coal trains passing through cities has given rise to local concerns about**
   A. noise pollution.
   B. reduced property values.
   C. interference with emergency transportation.
   D. All of the above

5. **In a coal-fired power plant, what must be done before the coal is actually burned?**
   A. It must be pulverized.
   B. It must be liquefied.
   C. The methane must be extracted from it.
   D. The oil must be extracted from it.

6. **If you want to use a backup generator to power your house during power black-outs, you must be certain to install and use**
   A. underground fuel tanks.
   B. an isolation switch.
   C. a device for burning off the exhaust.
   D. All of the above

7. In a combustion-turbine, oil-fired power plant, the electric generators are mechanically driven by
   A. steam turbines.
   B. gas turbines.
   C. hydro turbines.
   D. oil turbines.

8. In a coal-fired power plant, the electric generators are mechanically driven by
   A. steam turbines.
   B. gas turbines.
   C. hydro turbines.
   D. oil turbines.

9. Which of the following factors represents a significant disadvantage of methane compared with coal as a fuel for electric power generation?
   A. Methane burns less efficiently than coal does.
   B. Methane burns cooler than coal does.
   C. Methane causes more pollution than coal does.
   D. Methane is more dangerous to transport than coal is.

10. For the average homeowner, the inconvenience of an electric power blackout lasting a day or two can be completely eliminated in a practical fashion (without extreme overkill), allowing operation of all the home's appliances, by the use of a
   A. 40-MW coal-fired, on-site generator.
   B. 500-W gasoline-fueled, on-site generator.
   C. 20-kW propane-fueled, on-site generator.
   D. 60-MW oil-fired on-site generator.

chapter 11

Electricity from Water and Wind

Moving water and air are among the most natural forms of alternative energy. Nothing on this planet is truly infinite in supply, but the energy available from these sources, in practical terms, comes close to that ideal. Only solar energy is more nearly "eternal."

CHAPTER OBJECTIVES

In this chapter, you will

- Learn how large and medium-sized hydroelectric power plants work.
- See if you can install a small hydroelectric system for your home and expect it to work.
- Discover how ocean tides and waves can be harnessed to generate electricity.
- Learn how large-scale wind farms work.
- See if you can install a small wind turbine for your home and expect it to work.
- Compare stand-alone and grid-intertie small-scale hydro and wind power systems.

Large- and Medium-Scale Hydropower

A boom in hydroelectric power plant construction took place in the United States around the beginning of the 20th Century. As time passed, other types of power plants came online, including those fueled by coal, oil, methane, and nuclear fission. In America today, only a small fraction of electricity is generated by hydroelectric power plants, which can exist in three different forms. The best hydroelectric technology for a given location depends on the nature of the terrain, the present and anticipated future need for electricity, and the effects of the facility on plants and animals, water quality, agriculture, and overall quality of life.

Impoundment

An *impoundment hydroelectric power plant* consists of a dam and reservoir. This type of facility works best in mountainous places where high dams can be built and deep reservoirs can be maintained.

The potential energy available in a reservoir depends on the mass of water contained in it, as well as on the overall depth of the water. The potential energy in a specific parcel of water is expressed in *newton-meters* ($N \cdot m$). The *newton* is the standard unit of force, equivalent to a kilogram meter per second squared:

$$1 \, N = 1 \, kg \cdot m/s^2$$

Potential energy can be expressed as the product of the mass of the parcel in kilograms, the acceleration of gravity in meters per second squared (about $9.8 \, m/s^2$), and the elevation of the parcel in meters (the vertical distance it falls as its energy is harnessed). The equivalent kinetic-energy unit is the *joule* (J), which in effect equals a watt-second:

$$1 \, J = 1 \, W \cdot s$$

Figure 11-1 is a simplified functional diagram of an impoundment facility. The water from the reservoir passes through a large pipe called a *penstock*, and then through one or more water turbines that drive one or more electric generators. The output of each generator is stepped up in voltage and then sent to the transmission line for distribution.

Diversion

In a *diversion hydroelectric power plant*, a portion of the river is channeled through a canal or pipeline, and the current through this medium drives a

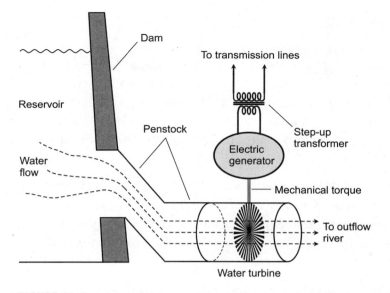

FIGURE 11-1 · Simplified functional diagram of a large generating system that derives its energy from water impoundment.

turbine. A dam is not required. This type of system is best suited for locations where a river drops considerably per unit horizontal distance. The ideal location is near a natural waterfall or rapids.

Small-scale and medium-scale diversion systems can be used next to mountain streams or other fast-moving, small rivers for the purpose of providing energy to individual homes or subdivisions. This type of system can be adapted to electric resistance heating systems, as we learned in Chap. 5.

TIP   *The chief advantage of a diversion system is the fact that, lacking a dam, it has far less impact on the environment than an impoundment facility.*

? Still Struggling

Obviously, a small-scale or medium-scale diversion system will not work if the stream or river dries up, or if it freezes solid all the way from the surface to the bottom!

Pumped-Storage

A *pumped-storage hydroelectric power plant* has two or more reservoirs at different elevations. When there's little demand for electricity among the consumers in the region served by the facility, the excess available power is used to pump water from the lower reservoir into the upper one(s). When demand increases, the potential energy stored in the upper reservoir(s) is released. Water flows out of the upper reservoir(s) in a controlled manner, passing through penstocks and turbines to generate electricity.

TIP *Pumped-storage systems require dams to hold the water in the reservoirs. These dams are generally smaller than the dams used in large impoundment facilities. Pumped-storage power plants can be found in regions where the terrain is hilly or gently rolling. But there must be a significant difference in average elevation between the reservoirs.*

Advantages of Hydroelectric Power Plants

- Hydroelectric power plants do not generate CO_2, CO, NO_x, SO_x, particulates, ground contamination, or waste products. Some heat goes into the stream or river water as a result of friction with the turbine components, but this is not always significant.
- Water is a renewable source of energy, as long as the stream or river does not dry up. The hydrologic cycle replenishes the source of potential energy in the form of rainfall, snowfall, and runoff.
- Hydroelectric power plant output can be controlled by changing the volume of water flow per unit time.
- The reservoirs created by an impoundment or pumped-storage power plant can be used for recreational purposes, and some people think that they provide dramatic scenery (although other people think that they ruin the existing scenery).
- The reservoirs created by impoundment are generally clean because impurities precipitate to the bottom. They can often be used as sources of water for drinking, bathing, washing, or irrigation.

Limitations of Hydroelectric Power Plants

- Large reservoirs flood land that could be used for something else. Whole towns have been sacrificed to reservoirs, causing displacement, resentment, and economic hardship.

- If a dam fails in a large impoundment facility, a catastrophic flood is almost certain to occur downstream.

- Hydroelectric power plants are not practical in regions where the terrain is flat.

- A prolonged drought can adversely impact, or even cut off, the energy production capacity of a hydroelectric power plant.

- In impoundment and pumped-storage power plants, the water level in the reservoir(s) varies considerably. One cannot expect to build a "beach house" directly on a reservoir!

- A dam can cause low dissolved-oxygen levels in the reservoir because it brings the normal river flow to a nearly complete halt. This phenomenon can kill fish and affect the nature of plant life in and around the reservoir.

- Dams can interfere with fish spawning. This problem can be mitigated by the use of *fish ladders*, *fish elevators*, or trapping and hauling the fish. However, such measures add to the cost of system construction or operation.

PROBLEM 11-1

In light of the problems with fossil fuels and the dangers of nuclear fission, why can't more hydroelectric power plants be built? There are plenty of rivers on all the continents of the world. Shouldn't we build as many hydroelectric facilities as we can?

SOLUTION

Most good sites for large hydroelectric power plants have been exploited already. A practical limit exists as to the number of dams and reservoirs that we can put along a river. Whenever energy is taken from a river by a hydroelectric facility, that energy is not available for use in any form further downstream. If too many hydroelectric power plants are placed along a river, energy availability and economic conflicts occur.

Small-Scale Hydropower

In *small-scale hydropower systems*, diversion technology is most commonly used, although impoundment can sometimes be done to a limited extent. A water turbine designed for home or small business use, installed in a fast-moving stream or small river with sufficient vertical drop, can produce 20 kW of electricity, more than enough for a typical household under conditions of peak demand.

How It Works

A small-scale hydropower system can be configured in three ways: stand-alone, interactive with batteries, and interactive without batteries. Interactive systems are also known as *intertie systems* or *grid-intertie systems*.

A stand-alone system employs banks of rechargeable batteries to store some or all of the electric energy produced by the water turbine. The batteries supplement the power from the turbine, and can provide all the electricity if the turbine ceases to produce power. An interactive system with batteries uses the electric utility, rather than the turbine, to keep the batteries charged. In an interactive system without batteries, excess energy is sold to the utility during times of minimum demand, and energy is bought from the utility during times of heavy demand.

Further details about how these three types of systems operate are provided in the section about small-scale wind power later in this chapter. The principles are basically the same for hydropower, wind power, and solar power.

TIP *Some states offer good buyback deals with utility companies. Other states do not. You should check the regulations in your state before you plan to install any sort of grid-intertie electrical system.*

Advantages of Small-Scale Hydropower

- Small-scale hydropower can reduce or eliminate dependence on conventional electric utilities.
- Water flows continuously if the stream is large and fast enough. It's more reliable than wind or solar energy sources, as long as the stream doesn't dry up or freeze solid from the surface to the bottom.
- Small-scale hydropower plants are virtually non-polluting. A small amount of heat is imparted to the stream water as a result of friction with the water turbine components, but this effect rarely causes any trouble.
- The electricity produced by a water turbine can be used for supplemental home heating or evaporative cooling.

Limitations of Small-Scale Hydropower

- Only a few people live on properties with streams having enough water flow to provide hydroelectric power.

- A small stream may periodically completely freeze or dry up, shutting a small-scale hydropower system down.

- A water turbine requires considerable water mass, along with a significant vertical drop, to provide enough power to heat a home. This may necessitate the installation of a small dam or artificial waterfall, which could give rise to environmental and regulatory issues.

- The up-front cost of a small-scale hydropower system is considerable. It takes a long time to pay for itself, and the resulting economic benefit may be outstripped by the initial cost.

PROBLEM 11-2

I'd like to install a stand-alone small-scale hydropower system for my home. I live on a ranch. A fairly good-sized stream runs through my property. An engineer has checked everything out. The vertical drop is sufficient, and there's more than enough water flow all year round. I'll need to build a small dam and back up some water to form a pond. That's okay with the local, state, and federal officials. But I'm concerned about how the system will affect wildlife.

SOLUTION

The wildlife-impact question is best answered by getting a wide variety of opinions. Naturalists from a nearby college or university can offer some insight. A pond can be expected to attract birds, fish, and other wildlife (some wanted, some not). However, the same pond will displace other wildlife, particularly mammals that dwell beneath the earth's surface.

Tidal-Electric Power

Ocean tides result from the gravitational pull of the moon and the sun, in conjunction with the rotation of the earth on its axis. In a simplistic sense, tides constitute massive oceanic waves with an extremely long *period* (time for the completion of a cycle), having two *crests* (high points) and two *troughs* (low points) each day in most locations.

The Tidal Barrage

A *tidal barrage* resembles a small dam with *sluice gates* that can be opened or closed, allowing water to flow between bodies of water having different elevations. This flow operates a water turbine that is mechanically coupled to an electric generator. The principle of operation is shown in Fig. 11-2.

As the tide comes in, the basin fills up through a large channel (not shown) until the tide reaches its highest point. The sluice gates are closed during this time, and the turbine does not operate. When the tide is at its peak, the elevation of the surface of the basin equals the level of the sea surface outside the basin. Then the tide begins to fall, and the basin acts as a reservoir. The sluice gates open so that water flows through the turbine, powering the generator. When the tide reaches its lowest point, water continues to flow through the turbine. But shortly after low tide, as the ocean level begins to rise again, the surface elevations of the basin and the ocean become equal. Then the sluice gates are closed, the turbine stops, and the basin fills up as the tide comes in once again.

An alternative system takes advantage of the incoming tide as well as the outgoing tide. Two basins are used, one constantly maintained at a level above that of the ocean and the other constantly maintained at a level below that of

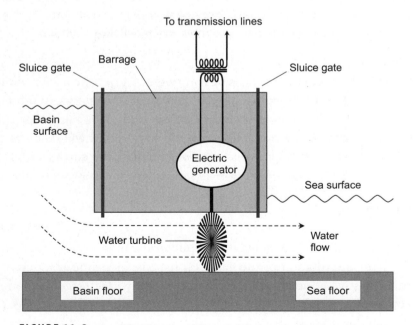

FIGURE 11-2 · Simplified functional diagram of a large generating system that derives its energy from a tidal barrage.

the ocean. The ideal "double-ended" tidal power plant of this sort has two systems like the one shown in Fig. 11-2, operating back-to-back with turbines facing in opposite directions.

TIP *A "double-ended" tidal-electric power system can theoretically produce twice as much energy, averaged over time, as a single-ended system can. But the "double-ended" system costs about twice as much to build as a "single-ended" system does.*

The Tidal Turbine

As the tides flow in and out in the vicinity of an irregular shoreline, currents appear in the water. *Littoral currents* flow parallel to the shoreline. *Rip currents* flow in large eddies near the shore. Tide-like currents also arise when large storm systems such as hurricanes pass.

Large-scale, persistent ocean currents prevail in various parts of the world. A good example is the *gulf stream* that flows eastward out of the Gulf of Mexico, around the tip of the Florida peninsula, and across the Atlantic Ocean toward the British Isles. Another example is the *Alaska current* that flows southeastward off the coasts of California, Oregon, and Washington State.

A *tidal turbine* bears a strong physical resemblance to a *wind turbine* (discussed later in this chapter). The turbine, and its associated support, are anchored to the ocean floor. Multiple turbines create a *tidal stream farm*. Each turbine is connected to an electric generator. The entire system lies beneath the surface, so it's invisible from above.

?

Still Struggling

Ocean currents travel more slowly than atmospheric winds, but water is hundreds of times more dense than air, so water produces much greater force per unit area on the turbine blades than air does. For this reason, tidal turbines are physically smaller than wind turbines.

Advantages of Tidal-Electric Power

- The tides are a renewable, reliable, and predictable resource.
- In a location with a significant difference between high and low tide, the ebb and flow can be harnessed to produce electrical power on a consistent basis.

- Tidal-electric power systems, like hydroelectric systems, generate no CO_2, CO, NO_x, SO_x, particulates, ground contamination, or waste products. Some heat is imparted to the ocean as a result of friction with turbine components, but it's rarely significant.

- Tidal-electric power systems seem exotic to some people. The existence of this type of facility in a given area can, therefore, be used to promote tourism, bringing in revenue.

- A tidal barrage can serve as a bridge for a roadway or railway across a bay or estuary.

- Maintenance of a tidal barrage is not difficult. The turbines last upwards of 30 years, and the barrage itself is inherently simple. However, the initial cost of installation is high.

- Tidal turbines are entirely beneath the surface. If installed in deep enough water, they present no obstacle or hazard to marine transportation.

Limitations of Tidal-Electric Power

- The installation of a tidal barrage is an expensive undertaking. Once in-stalled, however, maintenance is relatively easy.

- Tidal turbines can be difficult to install because the best sites for tidal currents are often in treacherous waters near rugged coastlines.

- Tidal-electric power plants can have a negative effect on marine life. Large fish, turtles, and marine mammals can be killed by the turbines, especially in tidal barrage systems. A large "catch" of this sort can also damage a turbine.

- A tidal barrage creates a sort of reservoir out of a bay or estuary, modifying its characteristics. This affects the *turbidity* (cloudiness) of the water and the extent of *sedimentation* (the settling of solid particles to the bottom).

- If not carefully designed and operated, a tidal barrage can cause localized flooding.

PROBLEM 11-3

Why are two systems of the type shown in Fig. 11-2, operating back-to-back, necessary in order to generate tidal power all the time? Can't a single-basin system be designed that takes advantage of the incoming tide as well as the outgoing tide, thereby generating power continuously?

✔ SOLUTION

A single-basin barrage system can be built that provides power *almost* all the time, but technical problems arise. Imagine a system with two sluices (call them the *inflow sluice* and the *outflow sluice*), each containing a turbine. Suppose that the system is designed so the elevation of the basin lags the elevation of the sea by one-quarter of a complete tidal cycle. When the level of the basin is higher than that of the sea, water passes through the outflow sluice, and the inflow sluice is closed. When the level of the basin is lower than that of the sea, water passes through the inflow sluice, and the outflow sluice is closed. Therefore, the basin and the sea have tides of equal magnitude but with different peak times, as shown in Fig. 11-3. The available power at any given moment in time depends on the difference in elevation between the sea and the basin, represented by the vertical distance between equal-time points on the curves. (At points where the curves intersect, no power is available because the levels of the sea and the basin are the same.) Unfortunately, this kind of system suffers from variability in power output. Moreover, it cannot provide any more total energy per unit tide cycle than a single-ended system. The most efficient and cost-effective way to get continuous energy from the tides is to employ two or more back-to-back systems similar to that diagrammed in Fig. 11-2, with independent basins and with their flow cycles timed so at least one of the systems always provides power.

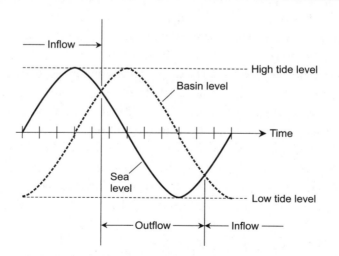

FIGURE 11-3 · A tidal barrage system in which the elevation cycle of the basin lags the elevation cycle of the sea by one-quarter of a tide cycle.

Wave-Electric Power

On large bodies of water, where the *fetch* is long (the wind blows over the surface for a great distance), friction between the air and the water surface causes *ripples* that grow into *wavelets*, then into *waves*, and ultimately into *swells*. Swells typically measure between one and three meters from crest to trough in the ocean near shore. Some swells, such as those generated by large and powerful hurricanes, can exceed 25 meters (80 feet) from crest to trough. Ocean swells contain the capacity to generate usable power. Normally this power dissipates when the swells reach the shore in the form of *breakers*. A *wave-electric generator* taps the power of ocean swells and converts it into electricity.

How It Works

Have you ever visited a swimming pool where artificial waves were made? This task can be done by pumping air in and out of a partially submerged chamber. If the chamber is airtight and watertight at all points that stay above the water surface (except for a hole where a pump is connected) but allows free flow at all points that stay below the surface, the water in the chamber cyclically falls and rises as air is pumped in and out of the hole. The wave maker converts the mechanical power from the air pump into power that propagates outward through the water.

Now imagine this process in reverse. Suppose that we place the same type of chamber a short distance offshore in a turbulent sea. This chamber, like the swimming-pool wave maker, has a hole above the surface. If we attach the chamber firmly to the bottom so that it can't move up and down, water swells will pass through it and cause the level of the water inside to rise and fall. The surface inside will remain essentially flat as its level fluctuates, so air will be pushed out of, and then drawn back into, the hole. The hole can contain an *air turbine* similar to the gas turbine in a jet engine. The turbine will rotate as the air is pumped in and out. The resulting mechanical torque can drive an electric generator. The output of the generator will go through a *power converter* that changes its electricity to 60-Hz AC (or 50-Hz AC in Europe) for distribution over electric transmission lines. Figure 11-4 is a functional diagram of such a system.

Advantages of Wave-Electric Power

- Ocean turbulence, like the wind and tides, constitutes a renewable resource.

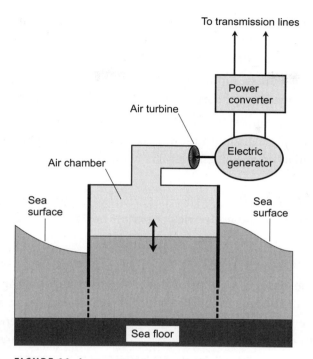

FIGURE 11-4 · Simplified functional diagram of a wave-power generating system.

- The conversion of wave power to electricity does not generate CO_2, CO, NO_x, SO_x, particulates, ground contamination, or waste products.
- A wave-electric generator is not particularly expensive to install or maintain, as long as it's built to withstand storms (without wasteful overengineering).
- Large "wave-power farms" can produce great quantities of usable electricity.
- Wave-electric generators have a low profile. Even when observed, they blend in fairly well with the scenery. (However, this asset can also manifest itself as a problem; note the fourth limitation below).
- Wave-electric generators, if properly designed, do not have a significant adverse effect on marine life.

Limitations of Wave-Electric Power

- When the ocean surface is calm or nearly calm, a wave-power generator will not produce usable output.

- Wave-electric generators must be sited carefully to minimize the effects of the noise they produce, but they must nevertheless be located where the energy from swells is available in sufficient amounts.

- A "hundred-year storm" might destroy a wave-electric generator unless it is overengineered to the extent that its cost does not justify its use.

- Wave-electric generators, because of their low profile, can present a hazard to marine navigation unless maps clearly show their locations. Buoys or other markers may be necessary.

PROBLEM 11-4

Are there any types of wave-electric generators besides the one diagrammed in Fig. 11-4?

SOLUTION

Yes. In one design, a string of floats, connected together by hinges, is placed on the water surface. The entire assembly undulates, and the resulting mechanical torque at the hinges can drive electric generators. Another design employs water turbines placed on the sea floor near a shoreline where strong *undertows* occur. An undertow is a current that flows away from shore after water hurled onshore by a breaker recedes down the sloping bottom. Other designs can take advantage of the pressure caused by breakers as they interact with seawalls or other obstructions placed at the shoreline.

Large-Scale Wind Power

The wind is one of the oldest sources of energy harnessed by humanity. *Windmills* have been used for centuries to pump water and to mill grain (that's where the "mill" in "windmill" comes from). Nowadays, wind power is making a comeback as problems with conventional energy sources increase.

Wind Speed, Operating Range, and Power

Wind speed can be specified in *meters per second* (m/s), *kilometers per hour* (km/h), *statute miles per hour* (mi/h), or *nautical miles per hour* (nmi/h), also known as *knots* (kt). The unit most often used by weather forecasters and other professionals is the knot, which equals approximately 1.852 km/h or 1.151 mi/h. The unit preferred by mass-media news and weather broadcasters in the United States is the statute mile per hour, which is equivalent to approximately 1.609 km/h or 0.8690 kt.

Manufacturers of *wind turbines*, especially in Europe, quantify wind speed in meters per second. If you know the wind speed in meters per second, multiply by 2.237 to determine the wind speed in miles per hour, or 1.943 to calculate the value in knots. If you're given the wind speed in miles per hour, divide by 2.237 to determine the speed in meters per second. If you know the wind speed in knots, divide by 1.943 to determine the speed in meters per second.

Most large wind turbines are designed to operate at wind speeds ranging from 3 or 4 m/s up to 20 or 25 m/s. That range corresponds to a minimum or *cut-in* wind speed of 7 to 9 mi/h, a breeze that lightly fans the face, and a maximum or *cut-out* wind speed of 45 to 56 mi/h, a gale against which you'll find it hard to walk. Within this operating range, a large wind turbine can generate anywhere from a few hundred kilowatts (kW) up to several megawatts (MW) of usable electric power, depending on the blade length, the wind speed, and the size of the generator.

Still Struggling

The *capacity factor* of wind energy—the proportion of time the resource can be exploited to produce usable output—is approximately 25 percent to 40 percent, depending on the geographic location and on the design of the turbine. The best sites for wind turbines are often far from population centers, necessitating the use of long transmission lines to get electricity to end users.

Design Considerations

Large wind turbines require tall, strong towers for support. A typical wind turbine tower measures between 50 m (165 ft) and 80 m (260 ft) high, and is anchored in a mass of concrete. Some towers are guyed for enhanced high-wind survival.

The blades in a large wind turbine range from approximately 25 m (80 ft) to 40 m (130 ft) in length, although some can get even longer than that. Most large wind turbines have three blades. Some designs have two blades, and a few variants have four or more. The *rotational diameter* equals twice the blade length. The blade bearing, electric generator, and generator cooling apparatus are contained in a housing called a *nacelle*. The system is designed to spin at a constant *angular speed* of approximately 20 revolutions per minute (rpm) for the entire workable range of wind speeds. Changes in wind speed within the workable range cause variations in the maximum deliverable output power, but not in the rate of blade rotation. A gear box translates the angular speed of the blades into the proper angular shaft speed for a generator to produce AC at a constant frequency of 60 Hz (in the United States and some other countries) or 50 Hz (in other parts of the world).

In order to function properly, a large wind turbine must be oriented so its blades rotate on an axis that points into the wind. Therefore, the plane defined by the rotating blades must remain perpendicular to the wind direction. To orient itself, the entire assembly (blades and nacelle) can swivel through a full 360° horizontal circle on a *turntable*. The wind direction and wind speed are detected by a *wind vane* and an *anemometer* similar to the instruments used by meteorologists for the same purposes. If the wind becomes too strong, a *fail-safe braking system* stops the blades and locks them in place, and the turntable rotates approximately 90° so that the assembly experiences the smallest possible wind load, minimizing the danger of damage.

In most large wind turbines, the rotor blades are on the windward side of the nacelle, as shown in Fig. 11-5A. This geometry is called *upwind design*. In some wind turbines, the rotor blades are on the leeward side of the nacelle

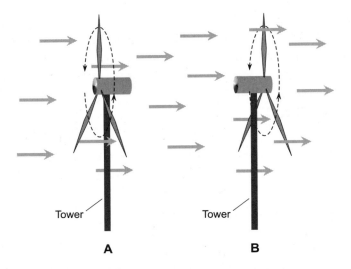

FIGURE 11-5 · At A, a large upwind turbine. At B, a large downwind turbine. Gray arrows represent the wind.

(Fig. 11-5B), an arrangement known as *downwind design*. Debate has taken place concerning which design produces more electrical energy in the long term for a given amount of money spent.

Proponents of the upwind design emphasize the fact that the technology is proven, and that significant improvements in efficiency and durability have taken place since the turn of the millennium. Those who favor downwind design argue that the blades can be designed to "fold" or "cone" away from the wind as they rotate, allowing the turbine to function at higher wind speeds than an upwind turbine can do, thereby increasing the capacity factor.

Figure 11-6 shows the basic components of a large-scale wind turbine connected to the utility grid. In some large-scale wind-power systems, groups of up to a half dozen turbines are connected in tandem, and the combination is connected to the grid. In the biggest wind farms, dozens of turbines are connected in tandem, and the combination feeds a cable that runs to a switchyard or high-tension transmission line.

Advantages of Large-Scale Wind Power

- Wind is a renewable energy source, and the supply is unlimited.
- Wind power plants do not produce greenhouse gases, CO, NO_x, SO_x, particulate pollutants, or waste products.
- Once installed, a large wind turbine is easy and inexpensive to maintain.

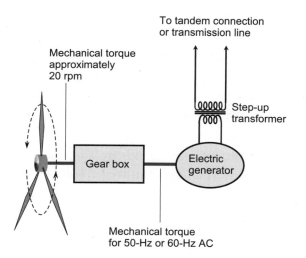

FIGURE 11-6 · Simplified functional diagram of a wind-power generator using a large turbine and transformer.

- Wind power plants can reduce dependency on fossil fuels, hydropower, and nuclear fission reactors for electric generation.
- Large wind turbines can be dispersed over wide regions, distributing the energy source and helping in the quest for a *fault-tolerant* utility grid (a system not vulnerable to catastrophic failure or sabotage).
- Wind power can supplement other modes of electric power generation, increasing the diversity of a nation's electrical system.
- Large wind turbines can be placed offshore over large lakes or over the ocean, as well as on land.
- Even the largest wind turbines have small footprints, and can thus share land resources with other operations such as farming and cattle ranching.

Limitations of Large-Scale Wind Power

- The wind is an intermittent source of energy. The capacity factor is lower than that of most other energy sources.
- A large wind turbine can be damaged or destroyed by a severe thunderstorm, hurricane, or ice storm.
- Some people dislike the physical appearance of large wind turbines, especially the distracting motion of massive rotors.
- Wind turbines make a certain amount of noise. However, at a reasonable distance from the tower, blade and turbine noise is hardly louder than the wind itself.
- Large wind turbines may occasionally injure or kill birds. This problem can be mitigated by judicious choice of location, and by not placing multiple turbines in close proximity with a common plane of blade rotation.
- Wind power cannot, by itself, satisfy all the electrical needs of a city, state, or nation. It is, at best, a supplemental source, used in conjunction with fossil fuels, nuclear fission, and hydropower.
- Locations with consistent usable winds are often far from population centers, requiring the use of long transmission lines.

PROBLEM 11-5

Why hasn't wind power been exploited to a greater extent? If wind turbines are distributed throughout the windiest regions of the United States, then there ought to be good winds blowing at some of the sites, no matter what the time. Wouldn't we have a constant supply of electricity if all the turbines were tied together into a single grid?

✔ **SOLUTION** _____

In theory, the answer to this question is "Yes." At any given time, there are plenty of locations in the United States where the wind is blowing at optimum speeds for the operation of wind turbines. The problem is finding a way to efficiently get the electricity from generating points to end users. Invariably, some end users live and work too far from operational generators to allow efficient transfer of the electricity to them.

Small-Scale Wind Power

The term *small-scale* applies to wind turbines that can generate as much as 20 kW of electricity under ideal conditions, enough to power most households. Like their larger counterparts, small-scale wind turbines generate power on an intermittent basis. In order to obtain a continuous supply of electricity with a small-scale wind-power plant, it is necessary to use storage batteries or an interconnection to the electric utility, or both.

How It Works

Most small-scale wind turbines are steered by a *wind vane* attached to the nacelle, rather than by a powered turntable. The vane works in the same way as an old-fashioned weather vane. When the wind blows hard enough to operate the turbine, the vane orients itself to point away from the wind, and the whole turbine resembles a miniature single-engine propeller aircraft without the wings. Under normal operating conditions, the plane defined by the blade rotation lies perpendicular (broadside) to the wind direction.

In a small-scale wind-power system, the speed of the blade rotation varies with the wind speed, resulting in variable-frequency AC from the generator inside the nacelle. This generator resembles the *alternator* in a motor vehicle. (Some manufacturers actually call it an alternator for that reason.) The AC from the generator is converted to DC by a *rectifier* circuit, and the DC charges a set of storage batteries. The electricity for household appliances comes from these batteries either directly, in which case special DC appliances must be used, or by means of a *power inverter* that converts the low-voltage DC electricity from the batteries to 117 V AC at 60 Hz (in the United States) or 50 Hz (in Europe and some other parts of the world).

When the wind speed exceeds a specified level, a small-scale wind turbine turns sideways to the wind to some extent. The plane defined by the blades is normally perpendicular to the axis of the vane. However, in a strong wind, the plane of the blades changes, so it no longer lies perpendicular to the vane axis. This geometric adjustment reduces the wind load on the blades but allows the turbine to keep on working. As the wind speed grows stronger yet, the angle between the plane of the blades and the vane axis decreases until, at a certain speed, it becomes zero. Then the blades rotate in a plane that contains the axis of wind flow. The variation in the angle between the plane of the blades and the wind direction is called *furling*. It can be done in the horizontal plane (so the blades swing, or *yaw*, toward the left or right) or in the vertical plane (so the blades tilt up or down).

Varying the *blade pitch* is another means by which a small-scale wind turbine can regulate its wind load. When the blade pitch is small (the plane of each blade's surface is nearly the same as the plane defined by the blades), the wind produces less force on the system, and consequently less power, than when the blade pitch is large (the plane of each blade's surface differs greatly from the plane defined by the blades). At low wind speeds, the blade pitch is maximum. As the wind speed increases, the blade pitch decreases. If the wind speed becomes great enough, the blade pitch becomes zero.

TIP   *In extremely high winds the blades can turn to zero pitch, furl completely, and lock in place. This function reduces the load on the blades as much as possible, minimizing the risk of structural damage. (It also shuts down the turbine.)*

Stand-Alone System

A *stand-alone small-scale wind-power system* employs rechargeable batteries to store the electric energy supplied by the rectified output of the generator. The batteries provide power to an inverter that produces a good AC wave at 117 V. In some cases the battery power is used directly, but this arrangement necessitates the use of home appliances designed for low-voltage DC. Figure 11-7 is a functional block diagram of a stand-alone small-scale wind-power system that can provide 117 V AC.

The use of batteries allows the system to produce usable power even if there's not enough, or too much, wind for the turbine to operate. A stand-alone system

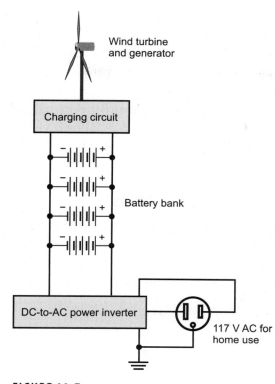

FIGURE 11-7 · A stand-alone small-scale wind-electric system.

offers independence from the utility companies. However, a blackout will occur if the system goes down for so long that the batteries discharge and no backup power source exists. This type of system is normally designed for a single home. Battery technology does not lend itself to large-scale wind-power systems.

Interactive System with Batteries

An *interactive small-scale wind-power-system with batteries* is similar to a stand-alone system, but with one significant addition. If there's a prolonged spell in which wind conditions are unfavorable for turbine operation, the electric utility can take over to keep the batteries charged and prevent a blackout. A switch, along with a battery-charge detection circuit, connects the batteries to the utility through a charger if no power issues from the turbine. When wind conditions become favorable and the turbine supplies power again, the switch disconnects the batteries from the utility charger and reconnects them to the turbine generator and rectifier.

Most interactive small-scale wind-power systems with batteries never sell any power to the electric utility, even if the wind turbine generates an excess. Power only flows one way, from the electric power line to the batteries through a charging circuit and switch, and even that happens only when the batteries require charging and the wind turbine does not provide enough power to charge them. Figure 11-8 is a functional block diagram of this type of wind-power system.

Interactive System without Batteries

An *interactive small-scale wind-power system without batteries* also operates in conjunction with the utility companies. Energy is sold to the companies during times of minimum demand, and is bought back from the companies during times of heavy demand. You can keep using electricity (by buying it directly

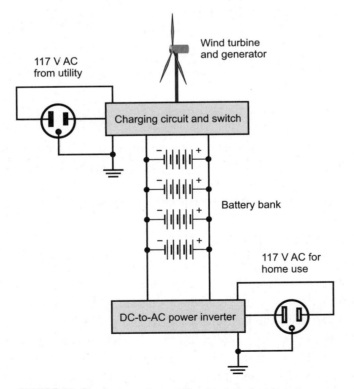

FIGURE 11-8 · An interactive small-scale wind-electric system with batteries.

from the utilities) if wind conditions become unfavorable for a prolonged period. Because this type of system has no batteries, it can be larger, in terms of peak power-delivering capability, than a stand-alone arrangement or an interactive system with batteries.

This type of system, like the interactive system with batteries, is designed to function with the help of the utility companies, and does not offer the independence that a purist might desire. This factor does not represent a technical drawback, but it can pose a philosophical problem for anyone who desires to live completely off the grid. Figure 11-9 is a functional block diagram of an interactive small-scale wind-power system without batteries.

Advantages of Small-Scale Wind Power

- Wind is a renewable energy source, and the supply is practically unlimited.
- Small wind turbines do not produce greenhouse gases, CO, NO_x, SO_x, particulate pollutants, or waste products.
- Once installed, a small wind turbine is easy and inexpensive to maintain.
- Small-scale wind power, when used in an interactive system, can reduce dependence on the electric utility.
- When used in conjunction with other alternative sources, small-scale wind power can offer independence from the utility.

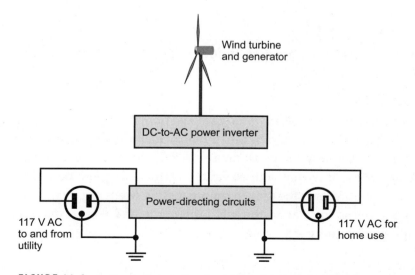

FIGURE 11-9 · An interactive small-scale wind-electric system without batteries.

- A small-scale wind power system with batteries can provide electricity in the event of a short-term utility blackout.
- Small-scale wind power can be used for supplemental home heating or cooling.

Limitations of Small-Scale Wind Power

- The wind is an intermittent source of energy.
- Even the best stand-alone small-scale wind-power system can provide only a small amount of electricity on a continuous basis.
- Small-scale wind turbines will not work properly if the wind is too strong.
- A small wind turbine can be wrecked by a powerful thunderstorm, hurricane, or ice storm.
- A small-scale wind-power system will take a long time to pay for itself, and in fact may never.
- Neighbors may dislike having a wind turbine nearby. (This attitude exhibits the "not in my back yard," or NIMBY, phenomenon.)
- Small-scale wind turbines can create significant noise at close range. They rotate at higher speeds than large turbines, and they usually sit atop relatively small towers, closer to the surface (and people's ears) than large-scale turbines.

PROBLEM 11-6

What factors affect the amount of power that can actually be derived from the wind?

SOLUTION

The available power P, in watts, available from the wind depends on the air density, the area defined by the turbine blades as they rotate, and the wind speed. If d represents the air density in kilograms per cubic meter (a value that varies somewhat with elevation, temperature, and barometric pressure), A represents the area in square meters swept out by the blades in a plane perpendicular to the wind direction, and v represents the wind speed in meters per second, then

$$P = dAv^3/2$$

In theory, therefore, the available power from a wind turbine varies in direct proportion to the air density, in direct proportion to the area swept out by the blades, and in direct proportion to the *cube* (third power) of the wind speed.

TIP   *If the wind speed doubles, say from 10 mi/h to 20 mi/h, then the theoretical wind power increases by a factor of 8 (or 2^3). If the wind speed triples, say from 10 mi/h to 30 mi/h, then the theoretical wind power becomes 27 (or 3^3) times as great. If the wind speed quadruples, say from 10 mi/h to 40 mi/h, then the theoretical wind power becomes 64 (or 4^3) times as great!*

? Still Struggling

With a conventional wind turbine, the available power from a wind of constant speed varies in direct proportion to the square of the blade length (that is, the square of the turning radius of the whole blade system). That's because the area of a circle (the shape of the region swept out by the blades as they rotate) is proportional to the square of its radius. This rule assumes that we don't change the number of blades or the general shape and pitch of each blade.

QUIZ

Refer to the text in this chapter if necessary. A good score is eight correct. You'll find the correct answers listed in the back of the book.

1. **We might expect a large-scale wind turbine to cut in at a wind speed of about**
   A. 2 mi/h.
   B. 8 mi/h.
   C. 20 mi/h.
   D. 50 mi/h.

2. **Potential energy, such as that available from impounded water, can be expressed in**
   A. meters per second squared.
   B. kilogram meters squared per second squared.
   C. watts.
   D. Any of the above

3. **Assuming a constant blade pitch and shape, we should expect that in a 20 mi/h wind, a three-blade wind turbine that describes a 10-foot-diameter circle would provide approximately**
   A. twice the power of a three-blade wind turbine that describes a 7-foot-diameter circle.
   B. three times the power of a three-blade wind turbine that describes a 7-foot-diameter circle.
   C. 1.4 times the power of a three-blade wind turbine that describes a 7-foot-diameter circle.
   D. 2.9 times the power of a three-blade wind turbine that describes a 7-foot-diameter circle.

4. **In order for a pumped-storage hydroelectric system to work,**
   A. a large waterfall must exist that never goes dry.
   B. the two reservoirs must be at essentially the same average elevation.
   C. there must be a significant difference in average elevation between the reservoirs.
   D. the water must have constant wave action.

5. **An impoundment hydroelectric power plant works best**
   A. in large lakes.
   B. on flat plains with shallow, wide, slow-moving rivers.
   C. in mountainous country.
   D. under the sea where currents prevail.

6. **A diversion type hydroelectric power plant works best**
    A. in a fast-moving river.
    B. under the sea where currents prevail.
    C. in an oceanfront location with large waves.
    D. in a slow, meandering river.

7. **In a location with substantial and reliable littoral currents, which type of hydro-electric system would work best?**
    A. Impoundment
    B. Diversion
    C. Reservoir
    D. Tidal

8. **A large wind turbine should turn at approximately**
    A. the same number of revolutions per minute as the wind blows in miles per hour.
    B. the same rotational speed throughout the entire working wind-speed range.
    C. 60 revolutions per second, so as to produce a good 60-Hz AC output wave.
    D. half the number of revolutions per minute as the blade length in feet.

9. **A stand-alone wind power system for a residential home must include**
    A. a device that allows for selling excess power to the electric utility.
    B. a storage battery to power the home when the wind does not blow.
    C. a device that allows for buying power from the utility when the wind does not blow.
    D. All of the above

10. **An ocean current can produce more power, in theory, than a wind at the same speed because**
    A. wind turbines are smaller than water turbines.
    B. wind turbines are less efficient than water turbines.
    C. wind turbines turn more slowly than water turbines.
    D. None of the above

chapter 12

Electricity from Atoms and Sunshine

Radiant energy can come from within the earth as well as from outer space. For the production of electricity, the most common terrestrial source of radiant energy is *uranium fission*, and the most common extraterrestrial source is the visible light (and to some extent IR and UV) that comes from the sun. *Hydrogen fusion*, while promising, is still in the research-and-development phase.

CHAPTER OBJECTIVES

In this chapter, you will

- Discover how atomic fission and fusion reactions can create energy.
- Learn how scientists have tried to harness hydrogen fusion to produce usable electric energy.
- Find out what a tokamak is, and see how it works.
- See how photovoltaic (solar) cells can be interconnected to form solar batteries.
- Compare stand-alone and interactive solar power systems.

Atoms

All matter is made up of countless tiny, dense particles. Matter is mostly empty space, but it seems "continuous" because the particles are submicroscopic and they move incredibly fast. Each chemical *element* has its own unique type of particle, known as its *atom*. The slightest change in the internal structure of an atom can make a tremendous difference in its outward behavior.

Protons, Neutrons, and Atomic Numbers

The part of an atom that gives an element its identity is the *nucleus*, which contains two types of particles, the *proton* and the *neutron*. A teaspoonful of either of these particles, packed tightly together, would weigh thousands of kilograms. Protons and neutrons have almost exactly the same mass, but the proton has a positive electric charge while the neutron has no electric charge.

The number of protons in an element's nucleus, the *atomic number*, gives that element its identity. The element with one proton in its nucleus is *hydrogen*; the element with two protons in the nucleus is *helium*. If there are three protons, we have *lithium*, a light metal that combines easily with many other elements. The element with four protons is *beryllium*, also a metal.

TIP   *In general, as the number of protons in an element's nucleus increases, the number of neutrons also increases, although some exceptions occur. Elements with high atomic numbers such as **uranium (atomic number 92), therefore, have far greater density than elements with low atomic numbers such as **carbon (atomic number 6).*

Isotopes

For any particular chemical element, the number of protons is the same in every atom, but the number of neutrons can vary. Differing numbers of neutrons result in different *isotopes* for a given element. All elements have at least two isotopes, and some have dozens.

Every element has one isotope that's most often found in nature. A change in the number of neutrons in the nucleus causes a change in the mass of an atom. Some isotopes stay the same over long periods of time; we call them *stable* isotopes. Other isotopes are not so well-behaved. Their nuclei fall apart, or *decay*. When atomic nuclei decay spontaneously, we call them *unstable*.

TIP *The nuclear decay process is always attended by the emission of energy and/or subatomic particles. These emissions are known as* radioactivity. *All the known isotopes of uranium are unstable. Some decay more rapidly than others.*

Atomic Weight

The *atomic weight* of an element is approximately equal to the sum of the number of protons and the number of neutrons in the nucleus. The most common naturally occurring isotope of carbon has an atomic weight of 12, and is called carbon-12 or C-12. The nucleus of a C-12 atom has six protons and six neutrons. Some atoms of carbon have eight neutrons in the nucleus rather than six. This type of carbon atom has an atomic weight of 14, and is known as carbon-14 or C-14. Uranium can exist in the form of U-234 (atomic weight 234) or U-235 (atomic weight 235) as well as the most common isotope, U-238 (atomic weight 238). In a nuclear power plant, U-235 is the isotope of choice because it has properties that make it possible to induce controlled decay, producing a steady supply of usable energy.

Ions

If an atom has more or fewer electrons than protons, then that atom acquires an electrical charge. A shortage of electrons results in positive charge; an excess of electrons gives a negative charge. The element's identity remains the same, no matter how great the excess or shortage of electrons. A charged atom is called an *ion*. When a substance contains many ions, the material is said to be *ionized*. Ionized materials generally conduct electricity well, even if the substance is normally not a good conductor.

?

Still Struggling

An element can be both an ion and an isotope different from the usual isotope. For example, an atom of carbon might have eight neutrons rather than the usual six, thus being the isotope C-14, and it might have been stripped of an electron, giving it a positive unit electric charge and making it an ion.

Power from Uranium Fission

The term *fission* means "splitting apart." In *nuclear fission*, the splitting-up of atomic nuclei produces smaller nuclei and different elements. This phenomenon can occur with many elements. In a nuclear power plant, nuclei of U-235 are split up deliberately in a regulated fashion. This process is called *induced fission*.

The Uranium Fission Process

The key to U-235 fission is bombardment by high-speed neutrons. When a neutron strikes a U-235 atomic nucleus, that nucleus splits almost instantly into two lighter nuclei. As this happens, two or three neutrons are emitted along with *gamma rays*, which are similar to *X rays* but more penetrating and energetic. Heat energy is also produced, warming up the uranium. If one of the emitted neutrons hits another U-235 nucleus, then that nucleus splits, and the process is repeated. In a large enough sample of U-235, the result of all this nucleus-splitting and internal neutron bombardment is a *chain reaction*. Eventually, if the chain reaction goes on long enough, all the U-235 ends up getting split down into nuclei of lighter elements. Then we're left with *spent nuclear fuel*.

One of three situations can prevail when U-235 is subjected to neutron bombardment. These scenarios are called the *subcritical state*, the *critical state*, and the *supercritical state*.

1. In the subcritical state, the reaction dies down before much fuel is spent. This happens if, on the average, less than one emitted neutron strikes another U-235 nucleus and splits it.

2. In the critical state, the reaction sustains itself in a steady fashion until the fuel is spent. This happens if, on the average, one emitted neutron strikes another U-235 nucleus and splits it.

3. In the supercritical state, the reaction increases in intensity, and the uranium heats up to the point that it melts. This happens if, on the average, more than one emitted neutron strikes another U-235 nucleus and splits it.

How a Fission Power Plant Works

In order for a nuclear fission reactor to work properly, it must be maintained in the critical state. This state of affairs involves controlling the temperature of the U-235, as well as starting out with a sample of U-235 having a certain mass and shape. To some extent, the amount of neutron radiation within the sample can be controlled to keep the system operating in a steady state. When controlled properly, a uranium fission reaction can provide large quantities of usable heat energy for long periods of time.

There's no risk of any "atomic explosion" with U-235 that has been refined specifically for use in nuclear reactors. If you've seen a movie in which a fission reactor blew up like a nuclear bomb, then that movie wasn't based on reality. Nevertheless, there's plenty to worry about if a fission reaction gets out of control. If a reactor is allowed to "go supercritical," the uranium will melt because of the excessive heat. That's a condition called *meltdown*.

TIP *Reactor-grade U-235 is not refined enough to undergo the sort of split-second, violent chain reaction that occurs in a nuclear bomb. However, a melt-down can contaminate the soil, water, and air with radioactive isotopes.*

A uranium fission reactor is housed in a multi-layered structure to keep radiation from escaping and to physically protect it from damage that could arise from external causes. A *radiation shield* and *containment vessel* prevent the escape of radiation or radioactive materials into the surrounding environment. The entire assembly is housed in a massive reinforced building called the *secondary containment structure*. This building, characteristically shaped like a dome or half-sphere for maximum structural integrity, is designed to withstand catastrophes, such as tornadoes, hurricanes, earthquakes, and direct hits by aircraft or missiles. But, once in awhile, such structures fail!

Figure 12-1 is a simplified functional diagram of a nuclear fission power plant. Heat from the reactor is transferred to a water boiler by means of heat-transfer fluid (coolant). The coolant passes from the shell of the boiler back to the reactor through the coolant pump. The water in the boiler gets converted to steam, which drives a turbine. The turbine rotates the shaft of an electric generator that is connected into the utility grid through a step-up transformer. After passing through the turbine, the steam condenses and goes back to the boiler by means of the feed pump. The water and heat-transfer fluid are in entirely separate, closed systems. Neither comes into direct physical contact with the other. This prevents the accidental discharge of radiation into the environment through the water/steam system.

Advantages of Fission Power Plants

- Uranium is a relatively inexpensive fuel. It can be found in regions widely scattered throughout the world.

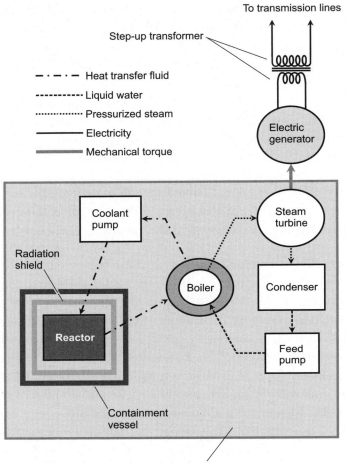

FIGURE 12-1 · Simplified functional diagram of a nuclear-fission power-generating system, showing one reactor and one turbine.

- Maintenance of fission power plants, while critical, need not be done as frequently as refueling and maintenance operations in conventional power plants.

- Fission reactors and their associated peripherals can operate in the absence of oxygen. That means they can be completely sealed and, if necessary, placed under the ground or under water without ventilating systems.

- Fission power plants do not produce greenhouse-gas emissions, CO gas, or particulate pollutants as do fossil-fuel power plants.

- Fission power plants, if responsibly built and used, can help the world economy wean itself off of its heavy reliance on fossil fuels for the generation of electricity.

Limitations of Fission Power Plants

- Uranium mining and refining can expose personnel to radioactive dust, and can also release this dust into the air and water.

- Fission reactors produce waste that remains radioactive for many years. Existing and proposed disposal processes for this waste are fraught with technical, environmental, and political problems.

- Transporting fissionable materials to power plants for use, and transporting waste products to nuclear dumps, can never be a perfectly secure business. The consequences of a security breach could be enormous.

- If certain fissionable nuclear waste products get into the wrong hands, nuclear terrorism or blackmail could result.

- Although the risk of accident or sabotage involving a nuclear reactor is small, the potential consequences—leakage of radioactive material into the environment—are great. The Fukushima Daiichi nuclear disaster, which took place in Japan after an earthquake and tsunami in March of 2011, serves as an excellent example of how bad things can actually get.

- The widespread use of fission reactors faces opposition from certain groups because of the above mentioned negative factors. This issue has given rise to public apprehension, particularly in Japan, concerning nuclear energy in general.

PROBLEM 12-1

What is the difference between *reactor-grade uranium* and *weapons-grade uranium*?

SOLUTION

For use in a fission reactor, uranium must be refined, or *enriched*, until it contains at least 3 percent U-235. In order to make fuel for an atom bomb, uranium must be enriched until it contains at least 90 percent U-235.

Power from Hydrogen Fusion

Because of public concerns about nuclear fission, hydrogen fusion has been suggested as a way to take advantage of the properties of the atom in order to generate electricity. In theory, that's a fabulous idea. Hydrogen fusion is more efficient in converting matter to energy than fission is, and no radioactive waste comes off. But a workable hydrogen-fusion reactor has not yet been developed.

Fusion in the Sun

Physicists and astronomers have determined that the sun converts hydrogen to helium by means of nuclear fusion. The term "fusion" means "combining." Hydrogen fusion requires extremely high temperature. The powerful gravitation imposed by the sun's huge mass keeps the core in a constantly compressed state. This compression keeps the core hot enough for hydrogen fusion to occur.

Solar hydrogen fusion is a multistep process, as shown in Fig. 12-2. At first, two hydrogen nuclei (protons, labeled H-1 in the drawing) combine, emitting a *positron*, also known as an *antielectron*. A positron has the same mass as an electron, but carries a unit positive charge rather than a unit negative charge. A *neutrino* is also emitted. Neutrinos have no electric charge, and can penetrate matter to an incredible extent, although they're harmless to humans. The fusing of two protons is attended by a loss of a unit positive charge; as a result, one of the protons becomes a neutron, producing a nucleus of *deuterium* (H-2), a heavy isotope of hydrogen comprising one proton and one neutron. The deuterium nucleus combines with another proton to form a nucleus of *helium-3* (He-3), containing two protons and one neutron. As this happens, a burst of gamma radiation is emitted. Two He-3 nuclei, resulting from two separate iterations of the above-described process, then combine to form a nucleus of *helium-4* (He-4), which has two protons and two neutrons. It's the isotope of helium we use to fill up lighter-than-air balloons. In this final phase, two protons come off. These particles can contribute to further fusion reactions.

In the solar fusion process, the total mass of the matter produced is a little less than the total mass of all the ingredients. The "missing mass" is converted into energy according to the so-called *Einstein equation*

$$E = m \, c^2$$

where E represents the energy in joules, m represents the "missing mass" in kilograms, and c represents the speed of light, equal to approximately 3×10^8 meters per second. The sun produces a tremendous amount of energy in this

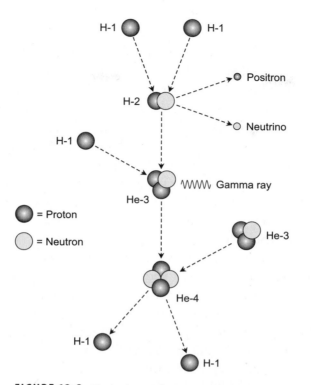

FIGURE 12-2 · The hydrogen fusion process that takes place in the core of the sun.

way because hydrogen nuclei "morph" into helium nuclei continuously and in vast numbers.

TIP   *Given any particular sample of matter, the Einstein equation tells us how much energy we could get, in theory, if we could completely convert the matter to energy.*

? Still Struggling

The sun contains enough hydrogen to keep its fusion process going for millions of centuries to come. Eventually the hydrogen fuel supply will run out and the sun will become unstable, but it won't happen soon enough for us to worry about it right now.

Fusion in Bombs

In a *hydrogen bomb*, a different hydrogen fusion reaction takes place. This mode, if it can ever be controlled, may function in a fusion reactor. Instead of simple hydrogen nuclei, which are protons, nuclei of *heavy hydrogen* merge. One nucleus is of deuterium (H-2), consisting of one proton and one neutron. The other nucleus is of *tritium* (H-3), which contains one proton and two neutrons. When these combine, the result is a nucleus of He-4, with the extra neutron ejected as shown in Fig. 12-3. Energy is also liberated. For this mode (called *deuterium-tritium fusion* or *D-T fusion*) to work, deuterium and tritium fuel must be supplied. Ordinary hydrogen (H-1) won't do the job. Several other fuel combinations can theoretically facilitate nuclear fusion, but the D-T mode has received the most attention.

In the sun, the fusion process goes on continuously because of the heat produced by the crushing pressure of gravitation, and also because of the heat generated from the fusion reactions themselves. In a hydrogen bomb, the necessary heat to start the reaction is supplied by a fission bomb, but the reaction burns itself out in a hurry. It's impossible to get enough gravitation to start and maintain fusion indefinitely and in a controlled manner in a terrestrial sample of simple hydrogen, deuterium, or tritium. In order to make hydrogen fusion generate useful power, we'll have to find a way to confine the fuel.

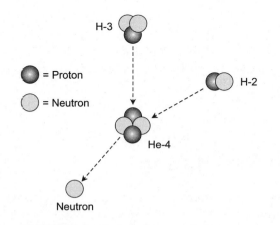

FIGURE 12-3 • The fusion process that occurs in a hydrogen bomb.

Plasma Fuel

When a sample of gas is heated to an extremely high temperature, the electrons are stripped away from the nuclei. The atoms, therefore, become ions. Instead of "orbiting" the positively charged nucleus of a specific atom, an electron is "free" to move from atom to atom, or even to travel through space all by itself. When this ionization happens, the gas, which is normally a poor conductor of electric current, becomes a good conductor. A substance in this state is known as a *plasma*. Because a plasma differs from an ordinary gas, the plasma state has been called the *fourth phase of matter* (the other three being solid, liquid, and gas). In the prototypes of fusion reactors that most scientists favor, deuterium and tritium exist in the plasma state, heated to temperatures comparable to those in the cores of stars such as our sun.

The behavior of a plasma can be dramatically affected by external electric or magnetic fields. An electric or magnetic field can cause a plasma to constrict, distort, bunch up, or spread out. If a plasma is surrounded by an external electric or magnetic field having certain properties, the plasma can be kept within a small, defined space, even if it becomes hot enough to sustain hydrogen fusion reactions. The external fields act on the plasma in much the same way as gravitation inside the sun keeps the hot core gases confined. The use of external magnetic fields to compress and hold a hot D-T plasma in place during a fusion reaction is known as *magnetic confinement*.

The Tokamak

One promising method of magnetic confinement makes use of an evacuated toroidal (donut-shaped) enclosure called a *tokamak*. This term is an acronym derived from a Russian descriptive phrase that translates as "toroidal chamber and magnetic coil." The plasma is contained inside the tokamak. Two sets of coils, called the *toroidal field coils* and the *poloidal field coils*, surround the toroidal enclosure (Fig. 12-4). The coils carry electric currents that produce strong magnetic fields. An electric current of several million amperes, provided by a large transformer, travels through the plasma around the toroid in a circular, endless loop. This plasma current creates a magnetic field of its own.

The magnetic fields from the currents in the coils and the plasma interact, confining the plasma, aligning it within the tokamak chamber, and forcing it toward the center of the chamber cross-section, keeping it away from the walls. This characteristic is important because the plasma must be heated to more than

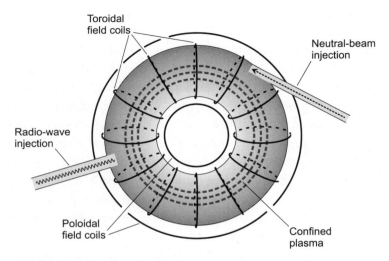

FIGURE 12-4 · Functional diagram of a tokamak, showing plasma confinement coils and two methods of heating the plasma.

100 million degrees Celsius (100,000,000°C) in order for fusion to occur! If the superheated plasma were to contact the tokamak wall at any point, the chamber would rupture, air would leak in, the plasma would cool below the critical temperature, and the fusion reaction would cease. The helium product of the fusion process is removed from the chamber by *diverters*.

Heating the Plasma

Several processes can be implemented in order to obtain the high plasma temperature necessary to sustain the fusion reaction:

- *Ohmic heating* arises from the fact that the current, as it circulates in the plasma, encounters a finite resistance. This means that a certain amount of power dissipates in the plasma, just as a wire gets hot when it carries high current.

- *Self heating* takes place because the fusion reaction produces heat itself, and some of this heat is absorbed by the plasma.

- *Neutral-beam injection* involves firing high-energy beams of neutral H-2 and H-3 atoms into the plasma. These high-speed particles heat the plasma when they collide with its atoms. The injected atoms must be electrically neutral so they can penetrate the powerful magnetic fields inside the tokamak. (Charged particles such as protons or helium nuclei would get deflected.)

- *Radio-wave injection* is done by transmitting electromagnetic (EM) waves into the plasma at several points in the chamber. These waves have a frequency such that their energy is absorbed by the plasma. Figure 12-4 shows one point of neutral-beam injection and one point of radio-wave injection.

Getting the Electricity

Figure 12-5 is a simplified functional diagram of a hydrogen fusion power plant. Except for the nature of the reaction, this type of power plant resembles a

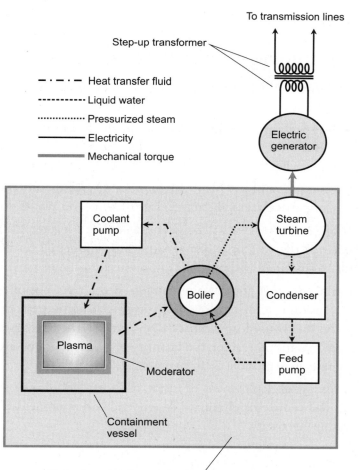

FIGURE 12-5 · Simplified functional diagram of a hydrogen-fusion power-generating system, showing one reactor and one turbine.

fission-based generating system. The plasma chamber, where the fusion reaction takes place, is surrounded by a *moderator*, which consists of lithium blankets that absorb neutron radiation from the fusion reaction. The high-speed neutrons cause the moderator to heat up, and also "breed" additional tritium fuel from the lithium.

Heat from the moderator is transferred to a water boiler, also called a *heat exchanger*, by means of coolant. The coolant is pumped from the shell of the boiler back to the plasma chamber. The water in the boiler is converted to steam, which drives a turbine. After passing through the turbine, the steam is condensed and sent back to the boiler by a feed pump. The turbine rotates the shaft of an electric generator that connects into the utility grid through a step-up transformer.

Advantages of Hydrogen Fusion for Power

- The only material by-products of hydrogen fusion are He-4, a harmless gas, and tritium that can serve as additional fuel.
- Deuterium fuel can be easily obtained from water. Lithium is abundant in the earth's crust. Tritium can be created in the reactor. These are the only material ingredients necessary to operate a D-T fusion reactor.
- A hydrogen fusion power plant will produce no greenhouse-gas emissions, CO gas, or particulate pollutants as fossil-fuel power plants do.
- A working fusion reactor will be safer than a fission reactor. Meltdown will not occur if the reactor is damaged because terrestrial fusion reactions cannot be sustained without the continuous infusion of fuel and energy.
- Terrestrial fusion is not a chain reaction. It can't get out of control and cause a fusion reactor to blow up. A hydrogen bomb explodes only because the fuel is used up almost instantaneously, not because of a chain reaction. In a fusion reactor, the quantity of fuel will not be great enough to generate an explosion.
- The widespread deployment of fusion reactors, should they ever be perfected, will reduce or eliminate dependence on nonrenewable fuels for generating electricity.

Limitations of Hydrogen Fusion for Power

- Although no radioactive waste is directly produced by D-T fusion, the emitted neutrons eventually make the reactor containment structure radioactive. This problem can be mitigated by using *low-activation* materials in the structure. Such materials become less radioactive from neutron bombardment than common containment materials such as steel. Unfortunately, low-activation alloys tend to be expensive.

- Although no radioactive waste is directly produced by D-T fusion, some radioactive tritium will be released by the reactor during normal operation. It has a *half-life* of 12 years, meaning that it takes 12 years to lose half of its radioactivity, 24 years to lose 3/4 of its radioactivity, 36 years to lose 7/8 of its radioactivity, and so on.

- The widespread deployment of working fusion reactors is not expected to take place until at least the middle of the twenty-first century. Major technical and logistic hurdles remain. In addition, the proponents of these systems will have to convince the public that hydrogen fusion reactors are safe.

PROBLEM 12-2

Will the powerful magnetic fields produced by the field coils and by the plasma pose a danger to the people who operate and maintain a magnetic-confinement fusion reactor?

✔SOLUTION

Not if the system is properly engineered. The containment vessel can be lined with a ferromagnetic metal such as iron or steel, which will keep the magnetic fields away from personnel in the vicinity.

Photovoltaics

A *photovoltaic* (PV) *cell* is a specialized form of *semiconductor diode* that converts visible light, IR radiation, or UV radiation directly into electricity. When used to obtain electricity from sunlight, this type of device is known as a *solar cell*. One of the most common types, the *silicon PV cell*, is made of specially treated silicon.

Structure and Operation

Figure 12-6 shows the basic structure of a silicon PV cell. It's made with two types of silicon, called *P type* and *N type*. The functional heart of the device is

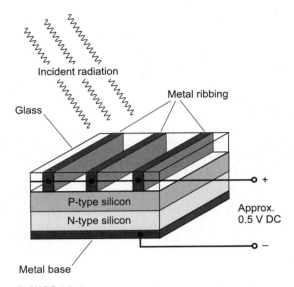

FIGURE 12-6 · Construction of a silicon photovoltaic (PV) cell.

the surface at which these two types of materials come together, known as the *P-N junction*. The top of the assembly is transparent so that light can fall directly on the junction. The positive electrode is made of metal strips or tiny bars called *ribbing* interconnected by fine wires. The negative electrode comprises a metal base called the *substrate*, which is placed in contact with the N type silicon.

When radiant energy strikes the P-N junction, a voltage or *potential difference* develops between the P type and the N type materials. When we connect a constant-resistance load to the cell, the intensity of the current through the load increases in proportion to the brightness of the light striking the P-N junction of the cell—up to a certain critical point. As the light intensity increases past this critical point, the current levels off at a maximum called the *saturation current*.

TIP *The ratio of the available electrical output power to the total visible-light power striking a PV cell (with both quantities expressed in the same units, such as watts) is called the* **efficiency**, *or the* **conversion efficiency**, *of the cell. The conversion efficiency can, and usually does, vary depending on how much visible-light power strikes the cell surface. We should, therefore, expect that a given PV cell will exhibit a different conversion efficiency in a dimly lit room than in direct sunlight.*

Voltage, Current, and Power

Most silicon solar cells provide about 0.5 V DC with no load connected. If there is no, or minimal, current demand, even fairly dim light can generate the full output voltage (symbolized V_{out}) from a PV cell. As the current demand increases, it takes more and more external illumination to produce the full V_{out}.

A maximum limit exists to the current that can be provided from a PV cell, no matter how intense the incident light might get. This specification is called the *maximum deliverable current*, and we symbolize it as I_{max}. The I_{max} value for a PV cell depends on the surface area of the P-N junction, and also on the particular technology involved in the manufacture of the device.

The maximum deliverable power (P_{max}) for a silicon PV cell, in watts, equals the product of V_{out} in volts and I_{max} in amperes. Because $V_{out} = 0.5$, we can write this fact mathematically as

$$P_{max} = 0.5\, I_{max}$$

? Still Struggling

In a PV battery made from two or more identical cells connected in series (negative-to-positive, like the links in a chain), the total voltage increases directly in proportion to the number of cells, but the maximum deliverable current remains the same as that of any individual cell all by itself. In a PV battery made from two or more identical cells connected in parallel (negative-to-negative and positive-to-positive, like the rungs in a ladder), the total voltage equals that of any cell taken alone, but the maximum deliverable current increases directly in proportion to the number of cells. When we combine series PV cells in parallel, or parallel PV cells in series, we can get more voltage and more current—the best of both scenarios. We call such a set a *series-parallel* combination of PV cells.

How Much Power?

The P_{max} of a series-parallel PV matrix made of identical cells equals the P_{max} of each cell, times the total number of cells. Alternatively, P_{max} equals the product of V_{out} and I_{max} for the complete matrix.

As an example, consider 10 parallel-connected sets of 36 series-connected PV cells, assuming $I_{max} = 2.2$ A for each series-connected set. We can calculate the following values for the matrix:

$$V_{out} = 36 \times 0.5 \text{ V}$$
$$= 18 \text{ V}$$

$$I_{max} = 10 \times 2.2 \text{ A}$$
$$= 22 \text{ A}$$

$$P_{max} = V_{out} I_{max}$$
$$= 18 \text{ V} \times 22 \text{ A}$$
$$= 396 \text{ W}$$

We can round this figure off to 400 W. It's a theoretical value, not a real-life result. When we connect a load to a PV system, V_{out} will turn out slightly lower than the theoretical amount. When numerous PV cells are connected in series, we observe a voltage drop of several percent under load as a result of the internal resistance of the combination. In the above case, V_{out} might be only 14 V when the current demand is near I_{max}. Therefore, the actual value of P_{max} would be

$$P_{max} = 14 \text{ V} \times 22 \text{ A}$$
$$= 308 \text{ W}$$

This value can be rounded off to 300 W.

Low, Medium, and High Voltage

In low-voltage, low-current PV systems, individual cells are normally connected in series to obtain the desired V_{out}. For charging a 12-V battery, a common V_{out} value is 16 V, requiring 32 cells in series. Such a series-connected set is called a *PV module*. In order to get higher I_{max}, multiple modules can be connected in parallel to form a *PV panel*. Finally, if even higher V_{out} or I_{max} is necessary, multiple panels can be connected in series or parallel to obtain a *PV array*.

Although high voltages (such as 5 kV DC) can theoretically be obtained by connecting hundreds or even thousands of PV cells in series, this approach presents problems because the *internal resistances* of the cells add up, severely limiting I_{max} and causing the output voltage to drop under load. This shortcoming can be overcome in high-power PV applications by connecting a large number of cells or low-voltage modules in parallel, making many identical such sets, and then connecting all the parallel sets in series.

If we want to get a medium voltage (say, the nominal voltage for a household utility circuit) from a low-voltage solar panel, we can use a power inverter along with a high-capacity rechargeable battery. The solar panel keeps the battery charged; the battery delivers high current on demand to the power inverter. Such a system provides common 117-V AC electricity from a 12-V DC or 24-V DC source.

Mounting and Location

A solar panel works best when it lies broadside to incident sunlight, so the sun's rays shine straight down on the surface. However, this orientation is not critical. Even at a slant of 45 degrees (45°) with respect to the sun's rays, a solar panel receives 71 percent as much energy per unit of surface area as it does when optimally aligned. Misalignment of up to 15° makes almost no difference.

Solar panels should be located where they will receive as much sunlight as possible, averaged out during the course of the day and the course of the year. Mountings should be sturdy enough so the panels will not rip loose or wiggle out of alignment in strong winds. One of the most popular arrangements involves mounting a solar panel, or a set of panels, directly on a steeply pitched roof that faces toward the equator.

The ideal bearing arrangement for a solar panel would be a motor-driven *equatorial mount*, similar to the ones used with high-end astronomical telescopes. This system would allow the panel to follow the sun all day, every day of the year. However, such a sophisticated mechanical device is impractical for most people, and the cost is prohibitive for large panels or multi-panel arrays.

The next best thing is a mount with a single bearing that allows for the panel to be manually tilted, always facing generally south in the northern hemisphere or generally north in the southern hemisphere. Figures 12-7 and 12-8 show this type of system for a single solar panel. Angles x, y, and z represent the tilt of the panel surface with respect to the zenith.

Northern Hemisphere

The arrangements in Fig. 12-7 will work between approximately 20° north latitude (20° N) and 60° north latitude (60° N). Cities in such locations include:

- Las Vegas, USA
- Chicago, USA
- Miami, USA

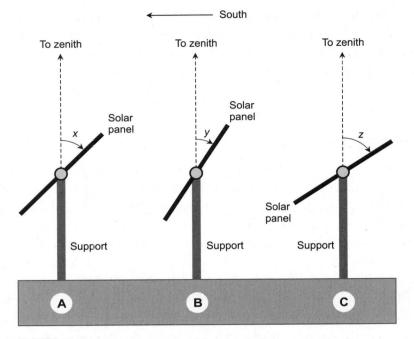

FIGURE 12-7 · Optimal placement of fixed, south-facing solar arrays for locations in northern temperate latitudes for (A) year-round operation, (B) low-solar-angle-season operation, and (C) high-solar-angle-season operation. The variables *x, y,* and *z* represent angles in degrees with respect to the zenith. In each case the panel is viewed edge-on, looking west.

- Paris, France
- Berlin, Germany
- Moscow, Russia
- Beijing, China
- Osaka, Japan

Figure 12-7A shows a year-round panel position. The angle x should be set to 90° minus the north latitude at which the system is located.

If an adjustable bearing is provided, two tilt settings can be used as shown in Figs. 12-7B and 12-7C. From late September through late March (autumn and winter), the arrangement shown at B is optimal, and the angle y should be set to approximately 78° minus the latitude. From late March through late September (spring and summer), the arrangement shown at C is optimal, and the angle z should be set to approximately 102° minus the latitude.

Southern Hemisphere

The arrangements in Fig. 12-8 will work between approximately 20° south latitude (20° S) and 60° south latitude (60° S). Cities in such locations include:

- Santiago, Chile
- Buenos Aires, Argentina
- Rio de Janeiro, Brazil
- Cape Town, South Africa
- Durban, South Africa
- Perth, Australia
- Sydney, Australia
- Auckland, New Zealand

Figure 12-8A shows a year-round panel position. The angle x should be set to 90° minus the south latitude at which the system is located.

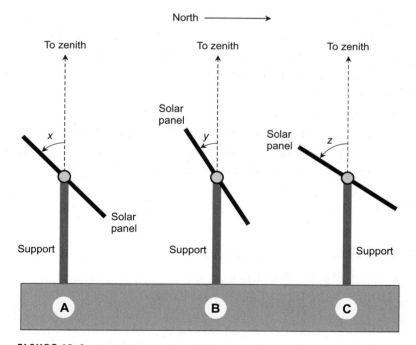

FIGURE 12-8 · Optimal placement of fixed, north-facing solar arrays for locations in southern temperate latitudes for (A) year-round operation, (B) low-solar-angle-season operation, and (C) high-solar-angle-season operation. The variables x, y, and z represent angles in degrees with respect to the zenith. In each case the panel is viewed edge-on, looking west.

If an adjustable bearing is provided, two tilt settings can be used as shown in Figs. 12-8B and 12-8C. From late March through late September (autumn and winter), the arrangement shown at B is optimal, and the angle y should be set to approximately 78° minus the latitude. From late September through late March (spring and summer), the arrangement shown at C is optimal, and the angle z should be set to approximately 102° minus the latitude.

PROBLEM 12-3

Suppose that a large number of solar modules are manufactured, each consisting of 53 identical PV cells in series. Each PV cell provides exactly 0.5 V at exactly 2 A in bright sunlight. You make two panels, each consisting of 20 of these modules in parallel. You connect these two panels in series. Suppose that the manufacturer of the modules tells you that under significant load, the output voltage will drop by 10 percent compared with the theoretical (no-load) value. What are the practical values of V_{out}, I_{max}, and P_{max} for this array?

SOLUTION

Let's calculate the *theoretical output voltage* (call it $V_{out\text{-}th}$) and then reduce this figure by 10 percent to get the actual output voltage, V_{out}. Then we'll calculate I_{max} for the whole array. Finally we'll determine P_{max} by finding the product of V_{out} and I_{max}.

Each module theoretically provides 53×0.5 V $= 26.5$ V. The parallel combination of 20 modules to form a panel also produces a theoretical output voltage of 26.5 V. (Identical voltages in parallel do not add up.) Two such panels are connected in series to form the array. Therefore

$$V_{out\text{-}th} = 2 \times 26.5 \text{ V}$$
$$= 53 \text{ V}$$

Reducing this value by 10 percent is the equivalent of multiplying it by 90 percent, or 0.9. Therefore

$$V_{out} = 0.9 \times 53 \text{ V}$$
$$= 47.7 \text{ V}$$

Each cell provides precisely 2 A of maximum deliverable current. Therefore, each series-connected set, forming a module, also provides 2 A of maximum deliverable current. (Identical currents in series do not add up.)

When 20 modules are connected in parallel to form a panel, the maximum deliverable current is 20 × 2 A = 40 A. When two of these panels are connected in series to form the final array, the resulting maximum deliverable current is still 40 A. Therefore, for the entire array, we have

$$I_{max} = 40 \text{ A}$$

In order to find the maximum deliverable power, we multiply V_{out} by I_{max}, as follows:

$$P_{max} = V_{out} I_{max}$$
$$= 47.7 \text{ V} \times 40 \text{ A}$$
$$= 1908 \text{ W}$$

This figure can be rounded off to 1900 W or 1.9 kW.

Large-Scale PV Systems

A large-scale PV system, designed to provide power to many users, is sometimes called a *solar farm*. The heart of this type of power plant is a massive array of PV cells. Solar farms can be found scattered around the sunniest parts of various nations. Depending on the size of the array, a solar farm can produce anywhere from a few dozen kilowatts up to a hundred megawatts (100 MW) or more.

How It Works

A large solar farm typically has thousands (and in some cases hundreds of thousands) of individual PV cells connected in a complex web of modules, panels, and arrays. In the most sophisticated large-scale solar farms, the arrays are set on equatorial mounts so that they can be turned directly toward the sun during all the hours of daylight. A computer-controlled mechanical system guides the bearings to optimize the orientation for every day of the year. The PV arrays connect to the utility grid through power inverters and transformers. Figure 12-9 shows the basic configuration.

TIP *In the biggest solar-electric farms, many inverters are connected in tandem, and their waves are maintained in exact phase coincidence. The combination feeds a step-up transformer that runs to a high-voltage AC transmission line.*

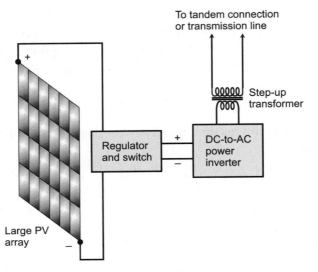

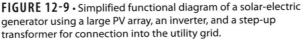

FIGURE 12-9 · Simplified functional diagram of a solar-electric generator using a large PV array, an inverter, and a step-up transformer for connection into the utility grid.

Advantages of Large-Scale PV Systems

- Sunlight is a renewable resource, and the supply is practically unlimited.
- Solar farms generate no greenhouse gases, toxic compounds, or particulate matter.
- In operation, PV cells make no noise.
- Solar farms, like wind farms, can be dispersed over wide regions. This approach distributes the energy source, an attribute that can help in the evolution of a fault-tolerant utility grid.
- Solar power can supplement other modes of electric power generation, thereby enhancing the diversity of a nation's electrical system.
- A typical solar farm has a low profile. No large towers or buildings mar the landscape.
- Solar farms constitute an important part of the long-term quest to minimize and ultimately eliminate humankind's dependence on nonrenewable fuels for generating electricity.

Limitations of Large-Scale PV Systems

- Large-scale PV systems are not cost-effective in places that get relatively little direct sunlight.

- Solar power, like wind power, is intermittent and cannot be stored on a large scale.

- Solar power cannot, by itself, totally satisfy the electrical needs of a city, state, or nation. At best, it's a supplemental power source.

- Unless proper electronic design precautions are taken and the site is chosen carefully, *load imbalance* can occur if part of a solar array is in sunlight while another part is in shadow. This situation reduces the output of the array more than would be expected by calculating the percentage of the surface area that lies in the shade.

- In order to get optimum performance from a large-scale PV system, the panels must be mounted on movable bearings. This arrangement can prove quite costly. A fixed-geometry system, no matter how well thought-out, is a compromise.

- Solar farms require a certain amount of dedicated real estate.

- Hail and wind storms can damage or destroy solar modules, panels, and arrays.

PROBLEM 12-4

What, exactly, is silicon, from which most PV cells are made? Is the supply of this material limited? Should we worry about running out of it someday?

SOLUTION

Silicon is an element with atomic number 14 and atomic weight 28. In the pure state, it's a lightweight, rather brittle metal, similar in appearance to aluminum. Silicon is a *semiconductor* substance, meaning that it conducts electricity very well under certain conditions and poorly under other conditions—and we can control those conditions. Silicon is abundant in the crust of the earth. There's enough to supply the needs of humanity indefinitely. In its natural state, silicon is almost always combined with oxygen or other elements. A good source is *silicon dioxide*, a major constituent of *quartzite sand*.

Small-Scale PV Systems

The term *small-scale* applies to PV systems that can produce enough electricity under ideal conditions to power a residential house. Like their larger counterparts, small-scale PV systems generate power on an intermittent basis. In order to obtain a continuous supply of electricity with a small-scale PV power plant, it is necessary to use storage batteries or an interconnection to the electric utility, or both.

Stand-Alone System

A *stand-alone small-scale PV system* employs rechargeable batteries to store the electric energy supplied by a PV panel or array. The batteries provide power to an inverter that produces 117 V AC. In some systems, the battery power can be used directly, but this method will work only with home appliances designed for low-voltage DC. Figure 12-10 is a functional block diagram of a stand-alone small-scale PV system that can provide 117 V AC for the operation of small household appliances.

The use of batteries allows the system to produce power even if there's not enough sunlight for the PV cells to operate. A stand-alone PV system of this type offers independence from the utility companies. However, a blackout will

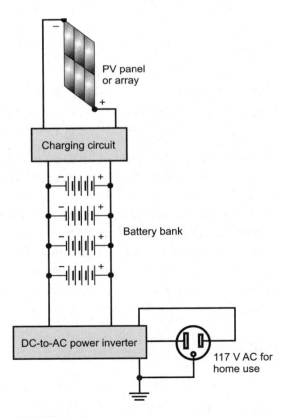

FIGURE 12-10 · A stand-alone small-scale PV system.

occur if the system goes down for so long that the batteries discharge and no backup power source is available.

Interactive System with Batteries

An *interactive small-scale PV system with batteries* is similar to a stand-alone system, but with one significant addition. If there's a prolonged spell without enough light for the PV cells to function, the electric utility can take over, keeping the batteries charged and preventing a blackout. A switch, along with a battery-charge detection circuit, connects the batteries to the utility through a charger if insufficient power, or no power at all, comes from the PV panel or array. When conditions become favorable and the PV cells can work again, the switch disconnects the batteries from the utility charger and reconnects them to the PV panel or array.

Figure 12-11 is a functional block diagram of an interactive small-scale PV power plant with batteries. In this system, power is not sold back to the electric

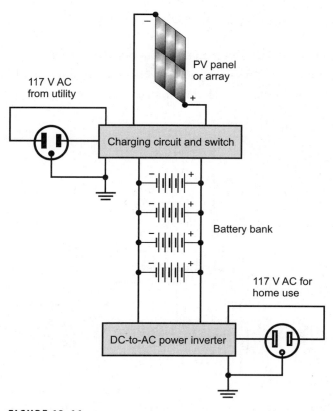

FIGURE 12-11 · An interactive small-scale PV system with batteries.

utility, even when the PV panel or array generates more power than the home needs. When the utility is involved, the electrical energy flows only one way, from the utility line to the batteries through a charging circuit and switch. That situation occurs only when the batteries require charging and the PV panel or array does not provide enough power to charge them.

Interactive System without Batteries

An *interactive small-scale PV system without batteries* operates in conjunction with the utility companies, just as the system with batteries does. Energy is sold to the companies during times of minimum demand, and is bought back from the companies during times of heavy demand. With this type of system, you can keep using electricity (by buying it directly from the utilities) if there's a long period of dark weather, and you don't have to "play nursemaid" to a set of batteries. Another advantage is that, because no batteries are used, this type of system can be larger, in terms of peak power delivering capability, than a stand-alone arrangement or an interactive system with batteries. Figure 12-12 is a functional block diagram of an interactive small-scale PV system without batteries.

TIP *In America, some states offer good buyback deals with the utility companies, and some states do not. Check the utility buyback laws in your state before investing in an interactive electric generating system of any kind.*

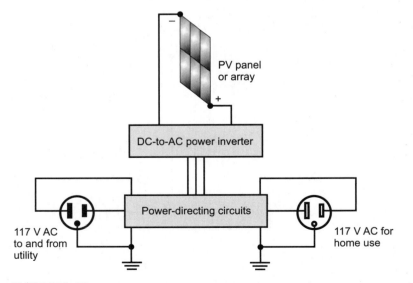

FIGURE 12-12 · An interactive small-scale PV system without batteries.

Advantages of Small-Scale PV Systems

- Sunlight is a renewable resource, and the supply is practically unlimited.
- Photovoltaic systems generate no greenhouse gases, toxic compounds, particulate matter, or other pollutants.
- In operation, PV cells make no noise.
- Small-scale PV systems can supplement other electric energy sources. The use of a small-scale PV system in conjunction with wind and/or small-scale hydropower makes it easier for a home to operate off the grid than reliance on a single alternative source.
- A typical small-scale PV system has a low profile. In some cases, it's hard to tell if it's there at all.
- Photovoltaic systems of all kinds are an important part of the long-term quest to reduce or eliminate dependence on nonrenewable fuels for generating electricity.
- A interactive PV system without batteries is simple. Not very many things can go wrong if the installation is done properly, and as long as reasonable care is taken to protect the solar panels from damage.
- There's practically no maintenance involved with a PV system that does not use batteries. Once it's up and running, you can pretty much forget about it.

Limitations of Small-Scale PV Systems

- Photovoltaics only provide power to a system when there's enough sunlight. Small-scale PV systems rarely justify the cost in places that get relatively little direct sunshine.
- If the solar panels get covered with snow or debris, you'll have to manually remove that stuff if you want the system to work.
- Problems with *load imbalance* can occur if part of a solar array lies in bright sunlight while another part lies in shadow.
- A hail or wind storm can damage or destroy a set of solar panels.
- In an off-the-grid system, you must make sure that the current demanded from the system does not exceed its maximum deliverable current.
- A small-scale PV system that can supply all the needs of a home (without any sacrifice or compromise) involves a large up-front cost that may never be recovered.

- In a PV system that uses lead-acid storage batteries, the batteries can produce dangerous fumes. They also require maintenance.

PROBLEM 12 - 5

In order to keep a stand-alone residential PV system running for prolonged cloudy periods, can't a huge storage battery bank be used—one so large that it can provide the necessary power for days or weeks without any charge from the PV panel or array?

✔ SOLUTION

In order for this scheme to work, you must find a way to ensure that the battery bank will attain a full charge from the PV cells after one or two sunny days. To do that, you'll need a gigantic PV array to match the massive battery. (A huge PV array could also provide limited charging current for the battery bank in gloomy weather.) The oversized battery bank and PV array will translate to an enormous up-front cost. Ventilating and maintaining the battery bank might prove problematic as well. Nevertheless, if you have unlimited financial resources, know some good engineers, have the required real estate for a big PV array, are determined to get off the electric utility grid, and live in a reasonably sunny location, then you can build a medium-scale, stand-alone PV system that will keep your house electrified all the time.

QUIZ

Refer to the text in this chapter if necessary. A good score is eight correct. You'll find the correct answers listed in the back of the book.

1. Consider a symmetrical or "square" series-parallel matrix of 25 solar cells (five series-connected sets of five cells each, all connected in parallel). Suppose that all the cells are identical, and each one can produce up to 0.40 watts of power in bright sunlight. Neglecting possible resistance losses, we should expect that the whole array will be able to deliver up to

    A. 0.40 watts.
    B. 2.0 watts.
    C. 4.0 watts.
    D. 10 watts.

2. According to the Einstein equation, we can get a certain theoretical amount of energy from any sample of matter. Imagine that we invent, and put into operation, a process to convert every single atom in any material object directly into energy. In that case, a 160-gram sample of matter will be able to produce

    A. the square root of 2 times the energy that an 80-gram sample can produce.
    B. twice the energy that an 80-gram sample can produce.
    C. four times the energy that an 80-gram sample can produce.
    D. eight times the energy that an 80-gram sample can produce.

3. Consider a symmetrical or "square" series-parallel matrix of 36 solar cells. Suppose that all the cells are identical, and each one can produce up to 1.5 amperes of current in bright sunlight. Neglecting possible resistance losses, we should expect that the whole array will be able to deliver up to

    A. 54 amperes.
    B. 9.0 amperes.
    C. 4.5 amperes.
    D. 1.5 amperes.

4. If each of the 36 solar cells in the situation of Question 3 can deliver up to 0.5 volt in bright sunlight, then (once again neglecting possible resistance losses) the entire array should be capable of delivering up to

    A. 18 volts.
    B. 9.0 volts.
    C. 4.5 volts.
    D. 3.0 volts.

5. As long as we express all energy or power figures in the same units throughout our calculations, we can define the conversion efficiency of a PV cell as

    A. the ratio of the maximum amount of energy that we can get from the cell in bright sunlight to the amount of energy we can get when we expose the cell to a one-watt source of visible-light energy.

B. the ratio of the mass of the cell to the amount of electrical energy we can get from it in bright sunlight.

C. the ratio of the total visible-light power striking the cell to the available electrical output power.

D. the ratio of the available electrical output power to the total visible-light power striking the cell.

6. **Scientists use the term "unstable" to describe atomic nuclei that**

A. accumulate neutrons easily.

B. increase in mass as time passes.

C. decay as time passes.

D. acquire and lose electrons easily.

7. **Imagine that you want to install a solar panel at your home in Fairbanks, Alaska. You look up the latitude of your town and find that it's 65 degrees north (to the nearest degree). If you use Fig. 12-13 as a guide to setting up your solar panel, optimized for year-round use, what angle q, as shown, will provide the best results?**

A. 25 degrees

B. 45 degrees

C. 65 degrees

D. Any angle between 25 and 65 degrees

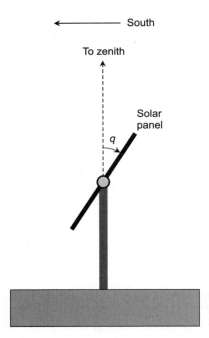

FIGURE 12-13 • Illustration for Quiz Question 7.

8. **The four phases of matter are**
   A. protons, neutrons, electrons, and energy.
   B. ions, isotopes, atoms, and molecules.
   C. solid, plasma, gas, and liquid.
   D. negatively charged, positively charged, neutral, and unstable.

9. **If an atom loses electrons, then**
   A. its electrical charge becomes more positive.
   B. it acquires an equal number of protons to maintain the same electrical charge.
   C. it becomes more likely to decay as time passes.
   D. All of the above

10. **Which of the following factors is an advantage of an interactive small-scale PV system without batteries?**
    A. You can sell excess energy to the utility company (in most places).
    B. The utility company can provide the energy you need when the sun doesn't shine.
    C. This type of system can deliver more peak power, in general, than a similar system that includes batteries.
    D. All of the above

chapter **13**

Illumination Technology

We consume a lot of electricity to light up our homes, businesses, streets, and general nighttime environments. In recent years, new technologies have emerged that provide more illumination using less energy than ever before. Let's compare the most common lighting technologies in use today.

CHAPTER OBJECTIVES

In this chapter, you will

- See how incandescent lamps work, and learn why some people want to phase them out.
- Learn how light dimmers work.
- Compare conventional fluorescent and neon lamps with incandescent lamps.
- Contrast compact fluorescent lamps (CFLs) with other lamp types.
- Contrast light-emitting diodes (LEDs) with other lamp types.

Incandescent Lamps

An *incandescent bulb*, also called an *incandescent lamp*, works by allowing an electric current to flow through a piece of wire that has a precisely tailored resistance and current-carrying capacity. As a result, the wire, called the *filament*, glows white hot. The bulb has a transparent, airtight case called the *envelope*.

Conventional Incandescent Lamp

The everyday incandescent lamp has a filament made of fine tungsten wire, which is often (but not always) wound into a "coiled coil" configuration. The manufacturer spins the wire into an extremely tight, long *helix* (coil); the resulting extended helical object is then wound into a larger helix. This geometry maximizes the length of wire that can fit inside the envelope, ensuring the greatest possible amount of energy emission in terms of the current provided. In other words, the filament design helps the lamp to work as efficiently as it can within practical constraints. The interior of the envelope comprises either a complete vacuum or a rarefied (low-pressure) inert gas. Figure 13-1 shows the basic construction details for a small incandescent lamp, such as the sort that you'll find in a battery-powered flashlight or lantern.

TIP *You can see the "coiled coil" filament geometry in almost any incandescent bulb that has a clear envelope. Examine the filament through a magnifying glass when it's not glowing (when the bulb is off).*

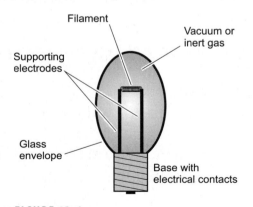

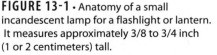

FIGURE 13-1 · Anatomy of a small incandescent lamp for a flashlight or lantern. It measures approximately 3/8 to 3/4 inch (1 or 2 centimeters) tall.

The current that flows through an incandescent bulb's filament depends on the voltage that you impose across it, and also on the end-to-end resistance of the filament wire. The filament resistance, in turn, depends on how hot it gets. A cold filament has lower resistance than a hot one. When you first apply voltage across the filament, a lot of current flows through it, so it warms up fast. As its temperature rises, the resistance increases, reducing the current, so the filament reaches a stable maximum temperature at which it glows white hot.

Incandescent lamps are notorious for low efficiency. The typical household incandescent lamp converts only about 10 percent of the applied power to visible-light output. The remaining 90 percent of the input power radiates from the device in the form of infrared (IR) rays, often called "heat." For this reason, energy-efficiency advocates have condemned incandescent lamps for many years. Nevertheless, many people keep using them because of their perceived "cozy" light output compared to more advanced lamp technologies.

? Still Struggling

The most common incandescent lamps are rated according to voltage (AC or DC, ranging from about 1.5 volts up to 240 volts) and the power they consume (from a fraction of a watt up to several hundred watts). In the United States, most DC incandescent lamps are designed for low voltage (from 1.5 volts up to 24 volts) while most AC incandescent bulbs are designed to work at 110 to 130 volts. Household incandescent lamps generally measure from 4 to 6 inches (10 to 15 centimeters) from the tip of the base to the top of the envelope.

Halogen Lamps

Traditional incandescent lamps suffer from short life spans as well as relative inefficiency. Tungsten gradually evaporates from the filament, limiting its useful life. The vaporized tungsten condenses on the inside of the glass envelope, gradually darkening it and further reducing the lamp's efficiency. These problems can be overcome, to a large extent, by filling the envelope with *halogen vapor* at low temperature. Iodine is the most common halogen used for this purpose, although chlorine and bromine have also been employed.

Figure 13-2 shows the internal anatomy of a small halogen lamp. The envelope is made of quartz rather than glass. A halogen lamp filament carries relatively more current, and therefore, operates at a higher temperature, than a conventional incandescent lamp does. Quartz can withstand higher temperatures than glass can tolerate. The higher filament temperature results in improved efficiency and "whiter" light at maximum current, compared with standard incandescent lamps. Halogen lamps are used in compact optical devices such as video projectors and desk lamps. They're also favored by the proprietors of art galleries, and they work quite well as automotive headlamps.

TIP   *Never touch the main (internal quartz) envelope of a halogen lamp when installing it, or when maintaining the supporting structure. If the quartz envelope is contaminated in any way, it will likely fracture when the bulb is illuminated to full or near-full brilliance.*

Dimmers

A *dimmer* is a simple electronic circuit that provides a variable effective output voltage when supplied with utility AC input. The device works by "chopping off" some of the AC wave to produce a lower effective voltage than the utility provides. A typical dimmer can handle fairly large loads in the form of multiple lamps of some (but not all) types. The dimmer consumes very little power itself.

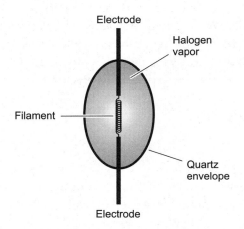

FIGURE 13-2 • Anatomy of a halogen bulb, such as might be used in a desk lamp. It measures approximately 3/4 inch (2 centimeters) tall.

The heart of a light dimmer is a component called a *silicon controlled rectifier* (SCR). The SCR, along with several simple *semiconductor diodes* and a *capacitor*, goes in the AC line to the lamps whose brightness levels you want to adjust. The amount of the AC wave that gets "chopped off" depends on the voltage at a so-called *gate* electrode. It works something like the valve in a water faucet. A variable resistor, called a *potentiometer*, allows adjustment of the control voltage at the gate.

A semiconductor light dimmer distorts the AC wave when set to any output voltage less than the full input (nominally 117 volts). The "chopping" of the wave produces abrupt current transitions every 1/120 or 1/240 second (twice or four times per AC cycle, depending on the circuit design). Because these transitions take place so fast, they generate *electromagnetic interference* (EMI) that can interfere with wireless devices, especially sensitive radio receivers.

TIP *The wiring in the voltage-controlled AC circuit acts like a transmitting antenna for the EMI. The greater the load controlled by the dimmer, the more severe this interference will likely be.*

Advantages of Incandescent Lamps

- Most people find the light from incandescent lamps esthetically pleasing.
- Incandescent lamps work well with light dimmers.
- Incandescent lamps (when operated without dimmers) don't produce electromagnetic noise, so they do not interfere with wireless devices.
- Incandescent lamps don't contain any toxic substances or environmental contaminants.

Limitations of Incandescent Lamps

- Incandescent lamps are inefficient; they consume far more energy than necessary to obtain a given amount of useful light.
- Halogen lamp envelopes can be weakened by contact with foreign substances, even the oil from your fingers.
- Incandescent lamps, and especially halogen lamps, can get extremely hot while in use. A high-wattage halogen lamp poses a fire hazard unless it's surrounded by a secondary envelope, a lens, or a protective grille.
- In some countries, governments have begun phasing out incandescent lamps, but high-wattage incandescent lamps may prove difficult to directly replace.

PROBLEM 13-1

What would happen to an incandescent lamp if the filament resistance decreased as its temperature increased?

✔SOLUTION

As the filament got hotter, it would draw more current from a constant-voltage source such as a lantern battery or AC utility outlet. The increased current would drive the temperature higher, which would cause the current to increase still more. Within a few seconds (maybe less than a second in smaller lamps), the filament would overheat and burn out. Engineers call that phenomenon, which occasionally occurs in poorly designed or malfunctioning electrical systems, *thermal runaway*. That's why, when incandescent lamps were in the invention phase many years ago, engineers chose filament materials whose resistance goes up with increasing temperature. That way, the filament acts as a current limiter, preventing thermal runaway.

Conventional Fluorescent Lamps

Fluorescent lamps produce light by exploiting the fact that electric currents through certain rarefied (low-pressure) gases cause the gases to glow. In addition, some solid and liquid substances will glow when struck by high-speed electrons, ultraviolet (UV) rays, or other high-energy rays. Fluorescent lamps are more efficient than incandescent lamps; they generate relatively more visible light and less IR.

Cylindrical Type

Until recently, the most common fluorescent lamp had a cylindrical geometry, sometimes bent into a ring or U-shape. Figure 13-3 is a simplified functional diagram of a fluorescent tube. During manufacture, the air in the sealed glass enclosure is evacuated, and then a small amount of mercury, sometimes along with an inert gas such as argon, is injected. The envelope is coated on the inside with a thin layer of phosphorescent solid material.

When an electric current is forced through the tube by applying a voltage between the end electrodes, the mercury turns to vapor and emits ultraviolet (UV) rays. The UV rays strike the phosphor, causing it to glow visibly white. Two phosphor types are commonly available: "warm white" (relatively more red) and "cool white" (relatively more blue). Fluorescent tubes are available for

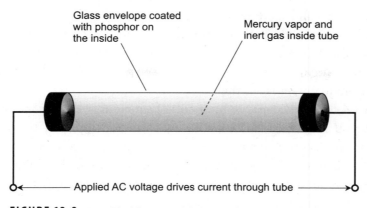

Glass envelope coated
with phosphor on
the inside

Mercury vapor and
inert gas inside tube

Applied AC voltage drives current through tube

FIGURE 13-3 · Simplified functional diagram of a cylindrical fluorescent lamp. It can range from a few inches to as much as 6 feet (about 2 meters) in length.

consumer use in various sizes and wattage ratings, from a few watts up to about 100 watts. Some specialized fluorescent tubes produce "colored" light.

A fluorescent lamp requires a "jump start" to vaporize the mercury before the current passing through the tube can make it glow. A device called a *starter* accomplishes this task. Some starters use a hot filament to vaporize the mercury; others cause a momentary high voltage to appear, causing a current surge that vaporizes the mercury.

TIP   *A common cylindrical fluorescent tube will glow in the presence of a powerful electromagnetic (EM) field, such as the sort that a radio transmitter produces. If you happen to be an amateur radio operator, try connecting the end of your antenna to a small fluorescent tube. Then make a transmission at 50 to 100 watts output, and watch the tube! Engineers, noticing this effect, have taken advantage of it to produce fluorescent lamps with exceptional efficiency by putting radio-frequency (RF) generators inside. Unfortunately, these fluorescent lamps emit radio waves that can interfere with wireless devices!*

Mercury- and Sodium-Vapor Types

When ionized, and when forced to carry high electrical currents, various elements produce emissions in the visible-light range. Mercury and sodium are especially useful for this purpose. Mercury vapor lamps emit white light while sodium-vapor lamps produce a yellow-orange glow with the color of a candle flame.

Large mercury- and sodium-vapor lamps are well known for their ability to produce extremely bright light. For this reason, they're good for outdoor lighting applications. Because mercury is toxic to humans and can accumulate in the environment, mercury-vapor lamps have passed from favor, and sodium-vapor lamps have gradually replaced them.

Figure 13-4 shows the basic components of a mercury-vapor or sodium-vapor lamp that you might find in a street light or yard light. In the mercury-vapor lamp, the interior of the envelope has a phosphorescent coating that glows white when the UV rays from the ionized mercury strike it. Both of these lamp types have far greater efficiency than incandescent lamps.

Neon Type

When confined at low pressure and forced to carry enough current to cause ionization, neon and argon gases emit visible light. Neon glows orange, while argon glows blue. Manufacturers have built electric lamps that work on this principle. The envelopes of these lamps don't need the phosphorescent inner coating that ordinary fluorescent lamps require, but neon and argon lamps produce far less light output than most other lamp types. (They also consume far less power!) Neon light technology isn't new; lamps of this sort were exhibited in a Paris automobile show as early as 1910. Artists have used neon and argon lamps to create exotic and spectacular exhibits.

You've seen neon signs if you've walked down a city's main thoroughfare at night. *Neon sign lamps* comprise long glass tubes filled with rarefied neon or

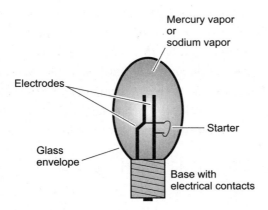

FIGURE 13-4 · Anatomy of a mercury-vapor or sodium-vapor lamp. It measures a few inches (roughly 15 to 30 centimeters) in height.

argon, bent into letters and numerals to form simple words such as "OPEN" or "SPECIAL." Neon sign lamps last for years. A smaller device of this type, the *neon glow lamp*, contains two electrodes in close proximity (Fig. 13-5). When you apply a sufficient DC voltage between the electrodes (usually on the order of 40 volts), a glowing region appears around the negative electrode. If you apply AC to the lamp, both electrodes seem to glow continuously, but the glow region actually shifts back and forth between the two electrodes as the voltage polarity alternates.

Advantages of Conventional Fluorescent Lamps

- Most fluorescent lamps are more efficient than incandescent lamps.
- Fluorescent lamps last longer, on the average, than incandescent lamps do.
- Even though fluorescent lamps cost more than incandescent lamps that give off the same amount of light, the improved efficiency over the course of the lamp's life will more than pay off the difference.
- Some people prefer the diffuse light given off by fluorescent lamps by virtue of their relatively large surface areas.

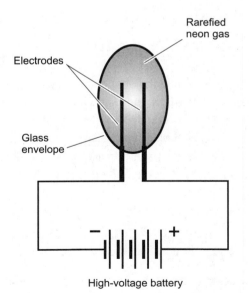

FIGURE 13-5 · Anatomy of a simple neon glow lamp with a DC voltage source.

Limitations of Conventional Fluorescent Lamps

- Some people dislike the spectral output of a typical fluorescent light; it can appear "harsh" or "too bright" or "too blue."

- Fluorescent lamps can malfunction and cause emission of electromagnetic noise that interferes with some radio reception, especially old-fashioned amplitude-modulated (AM) broadcast reception.

- Neon and argon lamps don't get very bright. You can't effectively use them to light up a room, for example, without installing so many of them that the cost becomes prohibitive.

- Some fluorescent lamps contain mercury, a known toxin.

- Fluorescent lamps do not work well in environments where they have to be switched on and off frequently. This activity shortens their useful life and reduces their efficiency as averaged over time.

- Conventional fluorescent lamps won't work with the dimmers that some people use with incandescent lighting systems.

?

Still Struggling

Have you ever wondered why a fluorescent lamp won't work with a common light dimmer? The main reason lies in the fluorescent lamp's so-called *ballast* device. The ballast provides a high voltage when the lamp is first switched on, so that the elements inside will vaporize and ionize quickly. A dimmer will prevent the ballast device from producing enough voltage to make the lamp start. Once the ionization has taken place, the ballast keeps the current stable so that the bulb doesn't "flicker." As the ballast device ages, it works less and less effectively; the bulb might start to "flicker" and never completely switch on, and ultimately it will fail altogether.

PROBLEM 13-2

You ask: "A friend once connected the sockets in a two-bulb table lamp in series with each other, rather than in parallel as the fixture was manufactured. She claimed that this wiring arrangement would make the bulbs last longer if two bulbs of the same wattage were installed. Will they really last longer? Can I do the same thing with fluorescent lamps? If I connect two of them in series with each other, will they both last longer than normal?"

✔**SOLUTION**

Fluorescent lamps don't tolerate unnaturally low voltages very well. When you connect two identical *incandescent* lamps in series and then plug the combination into a 117-volt utility outlet, each bulb gets half the voltage: 58.5 volts, more or less, depending on minor manufacturing differences between "identical" bulbs. They'll still work, although they will glow less brightly than normal. (And yes, they really will last longer!) However, conventional *fluorescent* lamp starters won't be able to generate sufficient voltage to properly ionize the elements inside the envelope, so this tactic won't work with them.

Compact Fluorescent Lamps (CFLs)

Fluorescent bulbs offer better efficiency than incandescent light bulbs do, but you can't usually remove an incandescent bulb from its socket and put a conventional fluorescent bulb in its place. The *compact fluorescent lamp* (CFL) was invented to satisfy people's desire for a direct-replacement fluorescent counterpart to the incandescent bulb.

Design Concept

Fluorescent and incandescent lamps differ in several ways. One of the most obvious, but also most easily overlooked, differences lies in the amount of light that the device produces *per unit surface area*. An incandescent lamp has a brilliant filament with a tiny surface area that emits all of the light, but the light from a fluorescent lamp comes off uniformly from a surface with considerable area. This difference presented the inventors of the CFL with a major challenge: How to make a fluorescent light bulb physically small without sacrificing overall brilliance.

A typical "old-fashioned" fluorescent tube has the shape of a long, thin cylinder. If the entire assembly is coiled up into a helix, the overall size of the lamp can be reduced. However, the amount of light emitted per square millimeter of bulb surface must increase if the lamp is to maintain its overall brilliance. The phosphor in a CFL must therefore produce far more visible light per unit area than the phosphor in a conventional fluorescent bulb does. For that reason,

CFLs burn hotter than conventional fluorescent bulbs do. The CFL phosphor must withstand the higher temperatures without rapidly degenerating.

The CFL concept originally came about during the "energy crisis" of the 1970s. Engineers at General Electric produced a working CFL model but decided against mass production because of the high startup costs involved. For consumer use, CFLs didn't gain much traction until the mid-1990s when China began exporting helical types. These CFLs had a harsh color similar to the output spectrum of conventional fluorescent lamps. By about the year 2005, this problem had been overcome. Today, you can get CFLs that produce light almost indistinguishable from the output of a typical incandescent bulb.

CFLs versus Incandescent Lamps

Today, if you see two otherwise identical table lamps with shades, one having a 60-watt "soft-white" incandescent bulb and the other having a CFL that produces the same amount of light output, you probably won't know which bulb is which until you look inside the shade. But a "60-watt-equivalent" CFL consumes only 10 to 12 watts of actual electrical power. You can also expect the CFL to last several times longer than an incandescent bulb does, although the CFL costs more at the store. Savings accrue over time as the result of less frequent bulb replacement and reduced overall energy consumption, so in the end, the CFL gives you the better deal.

One of the most significant problems with CFLs arises from the fact that they have trouble starting up in cold weather. If you live in a region where the winter temperatures drop below freezing, you won't like CFLs for outdoor use. They'll have trouble starting up, and once they do get started, they'll take a long time to reach full brightness. If the temperature gets extremely cold as it sometimes does in the northern USA and most of Canada, CFLs might not start up at all. Recent design improvements have mitigated this problem somewhat, but if you want to replace outdoor incandescent lamps with more efficient devices, you'll do better to go with *light-emitting-diode* (LED) lamps, described in the next section.

Another, less noticeable problem with conventional CFLs is a gradual decline in light output as they age. The newer designs suffer less from this trouble than the earliest ones did; some people don't even notice it until the old CFL burns out, and they install a new one of the same wattage. While you can rely on CFLs to last longer than incandescent lamps do, the lifespan difference is greatest in situations where you don't turn the lamp on and off very often. Yet another limitation of CFLs is the fact that the basic types won't work with dimmers. You can buy special "dimmable" CFLs, but they cost more than ordinary ones.

Radio-Frequency Interference

In some situations, CFLs exhibit a strange "quirk." The ballast produces high-frequency AC that can interfere with nearby wireless-device reception. Do you have a radio-controlled garage-door opener whose motor box has a light that comes on for a few minutes after you close the door? Is it an incandescent bulb? If so, try replacing it with a CFL, and see if you can still get the same operating range that you did before.

I discovered this *radio-frequency interference* (RFI) from a CFL by accident. The little remote-control device for my garage-door opener transmits a radio signal to the motor box. It normally works from up to 100 feet away. When I replaced the 60-watt incandescent lamp in the opener unit with a "60-watt-equivalent" CFL, the remote-control range went down to only about 20 feet. Because I'm an amateur radio operator (some people might even call me a "radio nerd"), I diagnosed the trouble right away. I put a new 60-watt incandescent lamp back in, and the remote-control box worked normally again.

TIP *If you notice performance issues with wireless devices (cordless phones, wireless routers, and the like) after replacing incandescent lamps with CFLs, try putting the incandescent lamps back into the fixtures one by one, and see if the problem goes away.*

? Still Struggling

You must use care when disposing of a burned-out CFL because it contains a small amount of mercury. Rules for disposal vary depending on where you live. If you break a CFL, you should avoid direct exposure to the contents when cleaning up the mess. You don't have to wear a hazmat suit, but you should wear a disposable dust mask to make sure that you don't inhale any of the airborne dust and debris. Those masks don't cost much; you can buy them at almost any department store or pharmacy. You should also wear disposable gloves (also cheap, and also available at department stores or pharmacies) and wash your hands after you've finished the cleanup task. Follow the instructions provided by the bulb manufacturer to prevent leakage of mercury into the environment, or call your local garbage removal service provider and ask them what to do. Demised-CFL disposal precautions are much the same as they are for automotive batteries that contain lead.

Advantages of CFLs

- CFLs are more efficient than incandescent lamps.
- CFLs last longer than incandescent lamps do.
- Over time, CFLs are cheaper to use than incandescent lamps.
- In some places, high-wattage incandescent bulbs are no longer sold, and CFLs offer a good choice for direct replacement.
- CFLs work well in situations where you must leave a light burning constantly for days, weeks, or months.

Limitations of CFLs

- CFLs lose a little of their brilliance as they age.
- If you break a CFL, you have to take special precautions to clean up the mess. Try hard not to drop one of them—and never simply toss them in the trash!
- If the temperature falls below freezing, CFLs won't start up very well. In extremely cold weather, CFLs won't work at all.
- CFLs can sometimes cause RFI to sensitive wireless devices nearby.

PROBLEM 13-3

Suppose that your regional utility company charges 10 cents per kilowatt hour (kWh) of electricity usage. If you replace two dozen 60-watt incandescent lamps with "60-watt-equivalent" CFLs that consume only 12 watts each, and if you burn each lamp for an average of 6 hours per day, how much money can you save over the course of a year, ignoring the up-front cost of the CFLs and the replacement costs of burned-out incandescent lamps? (Most incandescent lamps would likely burn out over the course of a year, but only a small fraction of the CFLs will.)

SOLUTION

Keep in mind the difference between energy and power! Watt-hours and kilowatt-hours quantify energy consumed over a specific period of time. Power in watts or kilowatts quantifies the rate at which energy is expended or used at some specific moment in time. Each CFL saves you $60 - 12 = 48$ watts of power compared to the incandescent lamp, at any given point in time. In 6 hours, that's 48×6, or 288 watt-hours (Wh) of consumed energy.

If you have two dozen bulbs, you save 288 × 24 = 6912 Wh of consumed energy in the 6 hours per day that the lamps burn. Remember that a kilowatt-hour equals 1000 watt hours; you therefore save 6912/1000 = 6.912 kWh a day. At 10 cents per kilowatt-hour, you save 69.12 cents ($0.6912) a day on the average. Over the course of a year, you will save $0.6912 × 365 = $252.29 (rounded off to the nearest penny). That sum will buy a fine supper for two in most American cities, or a great birthday party for your daughter or son!

Semiconductor Lamps

The term *semiconductor* arises from the ability of certain materials to conduct some of the time, but not all the time. Various elements, compounds, and mixtures can function as semiconductors. Common semiconducting materials include *silicon, gallium arsenide* (abbreviated GaAs), *germanium, selenium, cadmium* compounds, *indium* compounds, and the oxides of various metals.

Impurities

Impurities, also called *dopants*, give a semiconductor material the properties that it needs to function as an electronic component, and therefore, as a light-emitting or IR-emitting device. The impurities cause the material to conduct current under specific conditions. When manufacturers add an impurity to a semiconductor element, they call the process *doping*. The impurity material itself is called a *dopant*.

When a dopant contains an excess of electrons, we call it a *donor impurity*. A semiconductor material with a donor impurity is called *N type* because an electron carries a negative (N) electric charge. If an impurity has an inherent deficiency of electrons, we call it an *acceptor impurity*, and the resulting material conducts mainly by means of *hole flow*. A *hole* is a location within an atom where an electron "goes missing." A semiconductor material with an acceptor impurity is called *P type* because a hole has a positive (P) electric charge.

The P-N Junction

Connecting a piece of semiconducting material to a source of current can provide us with phenomena for scientific observations and experiments. But when the two types of material come into direct contact, the boundary between the

P type sample and the N type sample, called the *P-N junction*, behaves in ways that make semiconductors practical for producing visible light and IR.

When the N type material has a sufficient negative voltage with respect to the P type, electrons flow easily from N to P. The N type substance constantly "feeds" electrons to the P type substance in an "attempt" to create an electron balance, and the voltage source keeps "feeding holes" to the P type substance in order to sustain the electron imbalance. Figure 13-6A illustrates this condition, known as *forward bias*. Current can flow through a forward-biased diode easily. The voltage source "pushes" charge carriers toward the junction from both sides.

When we reverse the battery or DC power-supply polarity so that the N type material acquires a positive voltage with respect to the P type material, we have a condition called *reverse bias*, as shown in Fig. 13-6B. A semiconductor diode will not normally conduct when reverse-biased because an insulating *depletion region* forms around the P-N junction. The voltage source "pulls" the charge carriers away from the junction on both sides.

If you think of current as a movement of electrons (which, in physical reality, it is), then the diode shown in Figs. 13-6A and B can conduct electrons from

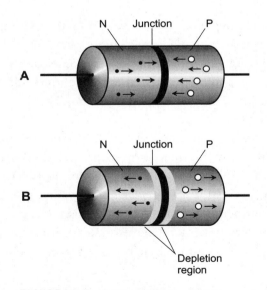

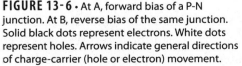

FIGURE 13-6 · At A, forward bias of a P-N junction. At B, reverse bias of the same junction. Solid black dots represent electrons. White dots represent holes. Arrows indicate general directions of charge-carrier (hole or electron) movement.

left to right, but not from right to left. The device, therefore, acts like a "one-way current gate."

Let the Light Shine!

Some diodes emit radiant energy when current passes through them in the forward direction. The current excites the atoms, causing the electrons to temporarily rise to unnaturally higher orbits around the atomic nuclei. As the saying goes, "What rises must fall." The electrons, left to their own ways, will inevitably fall back from these artificially high orbits into normal orbits. When the electrons fall back, they lose energy by radiating it in the form of visible light or IR.

Depending on the exact mixture of the materials used in their manufacture, *light-emitting diodes* (LEDs) can produce visible light of almost any color. An *infrared-emitting diode* (IRED) produces radiant energy at wavelengths slightly longer than those of visible red light.

The intensity of the radiant energy from an LED or IRED depends on the forward-going current, which in turn depends on the forward bias. As the current rises, the brightness increases, but only up to a certain point. If the current continues to rise, no further increase in brilliance takes place, and the LED or IRED then operates in a so-called state of *saturation*. If you drive too much current through the P-N junction, the device will burn out.

Digital Displays

Because LEDs can be produced in a great variety of shapes and sizes, they lend themselves to use in digital displays. You've seen digital clock radios, hi-fi radios, calculators, and car radios that contain LEDs. They make good indicators for "on/off," "a.m./p.m.," "battery low," and other conditions.

Some of the best flat-screen television (TV) sets employ LED technology. The screen can contain over a million individual diode elements, each of which contributes one *pixel* (picture element) to the display. An LED video display produces brilliant, true-color images, responds well to rapid changes in the input data (such as you get in action movies and computer games), and doesn't consume much energy.

If you're an environmental purist and you love large-screen TV sets, the LED display option should interest you!

TIP   *If you have a massive, bulky, old-fashioned* cathode-ray-tube *(CRT) TV set and you want to replace it with a flat-screen LED set, here's a good excuse to go* *out and fulfill your fantasy: Your new LED home entertainment unit will not only produce a better picture than your old set did, but the new set will cost less to operate in the long run. While large-screen LED TVs can set you back a lot of money up front, they will, in theory, last for many years with normal use and reasonable care. (Some manufacturers claim more than 100,000 hours of viewing.) You'll save hundreds, and maybe even thousands, of dollars on electric bills in that time.*

LEDs for Everyday Use

One of the most significant aspects of LED technology is the fact that it can produce a reasonable amount of light from only a small amount of electrical power. In fact, LEDs convert electrical power to visible light with even better efficiency than CFLs do. As a result, LEDs cost less to operate than any other type of lamp known (as of this writing). They produce less heat per watt of consumed energy than any other type of lamp, and they last longer. They don't contain toxic chemicals or generate any RF energy, as CFLs do. Most LEDs work okay with a light dimmer. Unfortunately, they're expensive. A single "40-watt-equivalent" LED will set you back about $30 (again, as of this writing). If you want to replace all of the incandescent lamps in your house with LEDs, you might spend upwards of $1000 straightaway on the conversion! In the long run, however, you'll save money because you'll hardly ever have to replace one of the things.

By their nature, LEDs are low-voltage devices, and they require DC to work. The 117-volt utility AC is entirely wrong for an LED; the voltage is too high, and AC is the wrong type of current. However, a simple semiconductor diode, acting as a "one-way current gate," can *rectify* the AC so it becomes *pulsating DC*. Series-connected LEDs (end-to-end, like the links in a chain) divide the high voltage equally among one another; so, for example, if you connect 20 LEDs in series across a nominal 60-volt pulsating DC source, each LED will receive 3 volts, which will do quite well.

The only trouble with the foregoing voltage-regulating arrangement lies in the fact that if a single LED in a series chain burns out, the whole set will go dark. Engineers get around this "bug" by connecting two or more LEDs in parallel (directly across each other, like the rungs in a ladder) for each spot in the series connection. That way, if a single LED fails, its parallel companion(s) can keep current flowing through the main series chain. Only the bad LED will go dark, instead of the entire congregation going out.

Although LEDs offer high efficiency in terms of their ability to convert electricity to visible light, no one has yet succeeded in designing an LED that shines with great brilliance. Nevertheless, LED lamps can be made bright enough to serve in many common devices including flashlights, lanterns, traffic lights, railway signals, electronic billboards, art galleries, and an increasing number of household light fixtures.

An LED's emitted-light color (that color that you actually see when you look at it from a distance) depends on the type of semiconductor materials used, and also on the presence of special internal phosphors that can change the color of the light that comes directly from the P-N junction inside the device. The earliest LED lamps had a characteristic gray-blue color that resembled daylight on an overcast afternoon. Nowadays, you can buy LEDs to suit almost any person's color wish.

Despite the promise of LED technology and the "cutting-edge" nature of a medium that appeals to technical and environmental purists, challenges remain before LEDs can be expected to predominate over other lamp types. The devices still cost a lot of money to manufacture, which translates to retail prices that many people don't believe they can afford. LEDs will not tolerate high temperatures. If you put LEDs in the ceiling fixtures of a hot room, or if you place LEDs in airtight fixtures, they'll probably fail as a result of the heat. The malfunction is, fortunately, not permanent. They'll work again after you switch them off for a half hour or so to let them cool down. But that sort of exercise doesn't really solve the problem, does it?

Advantages of LEDs

- LEDs last longer than other lamp types.
- LEDs are more efficient than other lamp types.
- Most LEDs work with light dimmers.
- LEDs don't contain toxic chemicals.
- LEDs don't produce RFI.

Limitations of LEDs

- LEDs have limited tolerance for heat.
- LEDs are relatively expensive to buy.
- LEDs have limited absolute brilliance.
- Some LEDs lose some brightness in the first few hundred hours of use.

? Still Struggling

A typical LED will work for about 30,000 hours if you operate it under reasonable conditions. According to that specification, if you burn an LED for an average of six hours a day, you can expect it to last for over 13 years. Of course, some LEDs will burn out sooner than that, and some will last longer.

PROBLEM 13-4

Once again, suppose that your electric company charges 10 cents per kilowatt hour. If you replace two dozen 60-watt incandescent lamps with "60-watt-equivalent" LEDs that consume 5 watts each, and if you burn the lamps for an average of 6 hours per day each, how much money can you save over the course of a year, ignoring the up-front cost of the LEDs and the replacement costs of burned-out incandescent lamps? (Most of the incandescent lamps would likely burn out in that time, but chances are good that you won't lose any of the LEDs.)

☑ SOLUTION

Each LED saves you $60 - 5 = 55$ watts compared to the incandescent lamp. In 6 hours, that's 55×6, or 330 Wh of energy. If you have two dozen bulbs, you save $330 \times 24 = 7920$ Wh, or 7.92 kWh, per day. At 10 cents a kilowatt-hour, you save 79.2 cents ($0.792) a day on the average. Over the course of a year, therefore, you'll save a grand total of $0.792 \times 365 = \$289.08$.

QUIZ

Refer to the text in this chapter if necessary. A good score is eight correct. You'll find the correct answers listed in the back of the book.

1. **The phosphorescent coating inside a mercury-vapor lamp**
    A. keeps the filament from overheating and burning out prematurely.
    B. allows the lamp to produce white light instead of the UV rays that the ionized mercury vapor itself emits.
    C. prevents the ionized mercury vapor from producing UV rays.
    D. ensures that all of the UV energy from the ionized mercury vapor can radiate from the lamp, maximizing the efficiency.

2. **If you need a bulb that will produce a lot of light, and the energy consumption or efficiency aren't important, you'd do best to choose**
    A. a halogen lamp.
    B. an LED.
    C. a neon lamp.
    D. Any of the above; it doesn't matter

3. **Simple semiconductor-based light dimmers work by**
    A. "chopping off" part of the AC wave to produce lower effective voltages than that provided by the utility.
    B. increasing the resistance across the devices connected to it, thereby reducing the current through them.
    C. increasing the temperatures of the lamp filaments, thereby making them draw less current.
    D. stepping down the effective voltage in exactly the same way as a utility transformer does.

4. **Which of the following lamp types is most likely to malfunction if operated in an extremely warm room?**
    A. Incandescent
    B. CFL
    C. LED
    D. Neon

5. **Which of the following lamp types is most likely to malfunction if operated outdoors on a cold winter night?**
    A. Incandescent
    B. CFL
    C. LED
    D. Neon

6. Which of the following lamp types is the most efficient? In other words, which one does the most effective job of converting electrical energy into visible light energy?

A. Incandescent
B. CFL
C. LED
D. Neon

7. The "coiled coil" geometry in some incandescent lamp filaments maximizes the

A. voltage through the wire.
B. current through the wire.
C. energy emission and efficiency.
D. current-to-voltage ratio.

8. An LED will glow when current flows through it

A. in the form of AC, but not in the form of DC.
B. in either direction.
C. in the reverse direction (that is, with reverse bias).
D. in the forward direction (that is, with forward bias).

9. Of the following lamp types, which one should you try hardest *not* to break because of the mercury it contains?

A. Conventional incandescent
B. CFL
C. LED
D. Halogen incandescent

10. The envelope of a conventional incandescent bulb contains

A. a pure reactive gas such as oxygen or chlorine.
B. air at high pressure.
C. air at normal pressure.
D. low-pressure inert gas or a complete vacuum.

chapter 14

Advanced Electrification Methods

We can generate electricity from the earth's interior heat, from the combustion or processing of biological waste matter, and from fuel cells. Electrical energy also exists in the form of a massive electrostatic charge between the earth and the upper atmosphere, courtesy of radiation from the sun.

CHAPTER OBJECTIVES

In this chapter, you will

- Learn how electricity can be generated from the earth's internal heat.
- Find out how engineers take advantage of plant and animal waste to generate useful energy.
- See how methane gas can be obtained from a compost heap.
- Compare fuel cells with other energy sources for obtaining electricity.
- Imagine how we might someday tap into the energy stored in the earth's atmosphere.

Geothermal Power

Translated from the Greek, *geothermal* means *earth heat*. The earth's core is incredibly hot, mainly as a result of the decay of radioactive materials with half lives measured in millions of years. In fact, if you go deep enough, the rocks in the earth exist in a liquid state called *magma*. Periodically this magma reaches the surface in various places. Then we observe volcanic eruptions.

Go Deep, Get Heat

On the average, the temperature of the earth increases by about 28°C for every kilometer, or 80°F for every mile, of depth below the surface for the first several kilometers down. In some locations, the temperature rises faster than that; in some locations it rises more slowly. Wherever you are, if you can sink a well deep enough, you can reach rocks that will boil water and get steam. That's the key to geothermal power generation.

The best locations for *geothermal power plants* exhibit subsurface temperatures that rise rapidly with increasing depth. Volcanic regions are great places in this respect! Geologically stable zones, or non-volcanic locations at high elevation, usually make poor geothermal-energy sites, but we occasionally find exceptions. Certain parts of the American West, for example, lie at high elevation, lack active volcanoes nearby, and nevertheless, offer promise for geothermal power generation.

Flash-Steam Systems

Figure 14-1 is a functional diagram of a *flash-steam geothermal power plant*. Water is forced down into an *injection well* by a *groundwater pump*. The well must go deep enough to reach subterranean rocks where the temperature is higher than the boiling point of water. The water filters through the rocks where it heats up and rises back up through the nearby *production well*. The hot water from the production well enters a *flash tank* where some of the water boils into vapor. Water that remains liquid in the flash tank goes back to the groundwater pump and gets forced down into the earth again.

The vapor from the flash tank drives a steam turbine, which turns the shaft of an electric generator. After passing through the turbine, the steam cools in a *condenser* so that the water returns to the liquid state and gets forced by the groundwater pump back down into the earth along with the diverted water

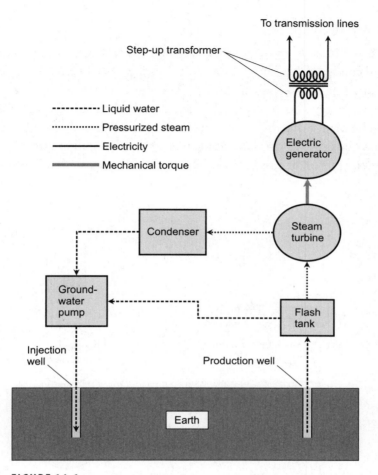

FIGURE 14-1 · Simplified functional diagram of a flash-steam geothermal power plant.

from the flash tank. Some of the condensed vapor can be used for drinking and irrigation because it is, in effect, distilled. The flash tank must be periodically flushed and cleaned to get rid of mineral buildup. If the water from the production well has high mineral content, the flushing must be done more often than if the water has low mineral content.

TIP   *In some locations, the subterranean rocks are so hot that the water from a geothermal power plant vaporizes on its way up through the production well. The flash tank is not necessary in this type of system, which engineers call a* **dry-steam geothermal power plant.**

Binary-Cycle Systems

In a *binary-cycle geothermal power plant* (Fig. 14-2), water is pumped into the earth and comes back up hot, just as it does in the flash-steam system. However, instead of going into a flash tank, the hot water enters a *heat exchanger* where most of its energy gets transferred to another fluid called a *binary liquid*. This fluid can be plain water, but more often it's a volatile liquid resembling a refrigerant that boils easily into vapor at a lower temperature than water. The liquid-to-vapor conversion occurs in a special low-temperature boiler. The pressurized

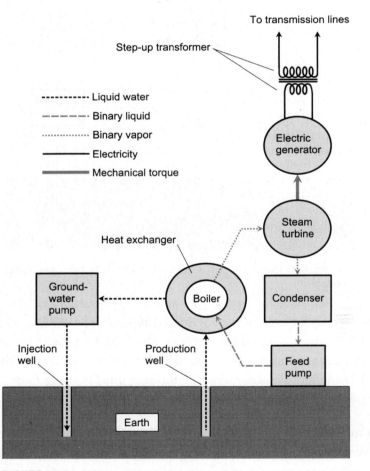

FIGURE 14-2 · Simplified functional diagram of a binary-cycle geothermal power plant.

vapor drives a steam turbine. Then the vapor leaves the turbine, gets cooled back into liquid by a condenser, and recirculates to the boiler.

The binary liquid remains in a closed system, isolated from the water that goes into the subterranean rocks. The binary-cycle system suffers from less mineral buildup than a flash-steam system does because none of the water that has passed through the rocks gets boiled off. In addition, no emissions escape into the atmosphere from a well-designed and wisely sited system. Binary-cycle geothermal power plants can sometimes work well in locations where the subterranean rocks aren't hot enough to operate a flash-steam or dry-steam system.

Advantages of Geothermal Power Plants

- The supply of geothermal energy is vast, although not infinite. We can consider it renewable, as long as excessive water is not pumped into the earth in one location in too short a time.

- A geothermal power plant does not need to have fuel transported or piped in from an outside source.

- The production of electricity from geothermal sources does not generate pollutants or toxic by-products. (However, see the last limitation below.)

- No external source of fuel is needed, except that required to initially start the pump(s). Once the power plant has begun operating, the electricity for the pumps can come from the plant's own output.

- After a geothermal power plant has been built, there are no operating costs except for routine maintenance and repair.

- A geothermal power plant has a low profile and does not take up a large amount of surface real estate.

- A flash-steam geothermal power plant, if placed on the shoreline of an ocean, can desalinate seawater for drinking and irrigation. This process results naturally from the distillation that occurs when the water boils to vapor and then recondenses.

Limitations of Geothermal Power Plants

- Finding a good site for a geothermal power plant, and getting approval from local residents or governments, can present a major challenge.

- In some cases an established geothermal power plant will "run cold." This phenomenon can occur as a result of natural changes in the subterranean environment. It can also occur if the site was poorly chosen and too much water is pumped down into the rocks.

- Flammable or toxic gases and minerals may be released from subterranean rocks and come up from the wells. These unwanted components can be difficult to get rid of. In some cases, they can be siphoned off and refined into fuel (crude oil and natural gas, for example). However, if the system is not well designed, gases, such as methane, may be released into the atmosphere.

PROBLEM 14-1

Can a geothermal power plant be built on a small scale to provide electricity for a single home or neighborhood?

✔ SOLUTION

Yes, sometimes—in locations where expensive, deep wells are not required. Perhaps the most noteworthy example is Iceland, which in effect sits on top of a huge volcano. Other possible locations include the areas near Yellowstone, Thermopolis, and Saratoga in Wyoming. The area around the town of Hot Springs in South Dakota presents another possibility.

Biomass Power

The term *biomass* describes a wide variety of plant and animal wastes. It literally means "biological matter." Biomass can take credit as one of the oldest sources of energy used by humankind, dating back to the discovery that wood produces useful heat when it burns.

Biomass Energy Sources

Biomass constitutes a renewable source of energy. It derives from the *photosynthesis* process, in which plants convert radiant energy from the sun into carbon-containing compounds known as *carbohydrates*. Plants, when grown specifically for use as biomass, act as a storage medium for solar energy.

When carbohydrates burn, they release heat, CO_2, and water. The CO_2 goes back into the environment and contributes to the *carbon cycle*, facilitating the growth of more plants to replace the spent biomass. Therefore, biomass can be CO_2-neutral if it is responsibly used. The water goes back into the *hydrologic cycle*. The heat energy can be harnessed for electric power generation, as well as for other human needs.

Some biomass, such as wood, can burn directly to release energy. However, various technologies have been developed that allow liquid and gaseous fuels to be derived from wood and other biomass substances. These fuels can supplement (and eventually perhaps replace) gasoline, petroleum diesel, methane, and propane. The following are several common raw materials for biomass energy systems:

- Trees and grasses: Wood and grass can be directly burned to provide heat for boilers, which drive steam turbines. The most common source of wood biomass is the waste from lumber and paper mills. Willow trees, switchgrass, and elephant grass are grown especially as biomass for energy production.

- Crops and crop residues: Corn is used to make ethanol. The same fact holds true, to a lesser extent for grains, such as wheat, rye, and rice. Sugar cane is used in Brazil to produce ethanol. Soybeans, peanuts, and sunflower seeds have been "refined" to make biodiesel fuel. Both of these fuels can serve to generate electricity or to propel motor vehicles.

- Aquatic and marine plants: Microalgae, found in certain lakes, can be fermented to obtain ethanol or composted to obtain *biogas* (biologically produced methane). Ordinary seaweed can also be harvested for this purpose.

- Manure and sewage: Animal waste from farms and ranches, and also human waste from urban areas, can be added to compost piles to accelerate the generation of biogas.

- Landfills: Many types of ordinary garbage, particularly paper, cardboard, and discarded food products, can be composted to obtain biogas.

A Biogas Example

The composting of plant and animal waste can produce combustible methane gas. Have you heard of the "swamp gas" that can accumulate in wetlands and sometimes catch fire? That's biogas, courtesy of Mother Nature! It's the same fuel as the commercially or privately produced methane that can facilitate home heating, vehicle propulsion, and utility electrification.

The flowchart of Fig. 14-3 shows how methane gas can derive from the composting of plant and animal waste in dedicated facilities for electric power generation and other purposes. Figure 14-4 is a functional diagram of a combined-cycle, methane-fired power plant that derives its fuel from the composting of biomass on-site.

Advantages of Biomass Power Plants

- Biomass is a renewable energy resource.

- Biomass power, if responsibly used, produces no net CO_2 emission because the new fuel grown absorbs all the CO_2 generated by the spent fuel.

- Biomass fuel does not produce very much sulfur-based pollution (SO_x), even when it is directly burned. In general, the SO_x production is less with biomass fuels than with conventional fossil fuels.

- Large biomass power plants can operate on a continuous basis, unlike solar and wind power plants that produce energy only when the sun shines or the wind blows.

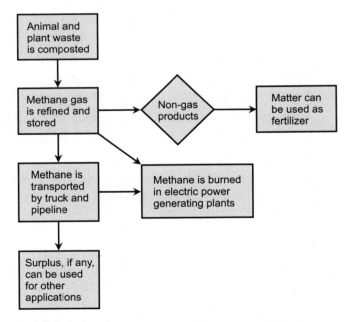

FIGURE 14-3 · Methane gas can be produced by the composting of plant and animal waste in dedicated facilities.

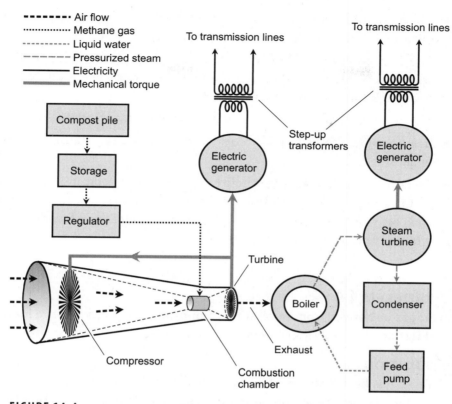

FIGURE 14-4 · A combined-cycle methane-fired power plant that derives its fuel from the composting of biomass onsite.

- Methane can be produced in small-scale composting plants. The supply does not have to come exclusively from centralized sources. Diversified composting could enhance the security of the civilized world by distributing energy resources and assets, making them less vulnerable to natural or human-caused disasters.

- Some of the plants used for biomass power, such as switchgrass, can reduce erosion of the landscape and provide a habitat for wildlife.

Limitations of Biomass Power Plants

- Biomass combustion generates some pollutants. The nature of the pollutants depends on the fuel burned. Nitrogen oxides (NO_x) are fairly common. Burning plant matter directly can generate significant CO and particulate pollution.

- Collection of matter for biomass power plants can impact the environment in adverse ways, if not responsibly done.
- The transportation of biomatter to composting plants, or to facilities where they are burned directly, consumes energy, usually in the form of fossil fuels for trucks and trains.
- The production of biogas by composting can produce objectionable odors.
- Irresponsible production of biogas can breed disease-causing microorganisms.
- Tanks or other containers that hold biogas require periodic inspection and certification by licensed and qualified personnel. This can be inconvenient and costly, but it is an absolute requirement to ensure the safety of people who live and work near the system.

PROBLEM 14-2

Can biogas be derived from small-scale composting and stored on-site for the electrification of a single home or neighborhood?

SOLUTION

Yes. It has been done using backyard compost heaps, makeshift storage containers, and small methane-powered electric generators. However, odor problems often arise. A more serious concern is the on-site storage of methane gas, which can present a fire or explosion hazard. Before anyone sets up such a system, they should check the local zoning and fire ordinances and abide by all applicable regulations. After the system has been built, a qualified civil engineer will have to regularly inspect it to ensure that it remains safe to operate.

Small-Scale Fuel-Cell Power

A fuel cell converts combustible gaseous or liquid fuel into usable electricity, but at a lower temperature than normal combustion does. In practice, a fuel cell behaves like a battery that we can recharge by filling a fuel tank, or if the fuel is piped in, by a continuous external supply. In Chapter 8, we learned the basics of fuel-cell operation, and how these devices can facilitate vehicle propulsion. The same technology can apply to small-scale electrification.

How It Works

Figure 14-5 is a functional block diagram of a small-scale, fuel-cell power plant suitable for a home or small business. This system can also work for recreational vehicles (RVs) and boats. In the case of a fixed land location, the fuel can be stored on-site or piped in. Engineers have suggested conventional methane as an ideal fuel source for home power plants of this type because the delivery infrastructure exists right now, and on-site storage is not necessary. However, in rural areas, or in any location not served by methane pipelines, other fuels might prove more cost-effective.

A typical fuel-cell stack delivers several volts DC, comparable to the voltage produced by a solar array or automotive battery. Under normal conditions, the DC from the fuel cell goes to a power inverter that produces usable 117 V AC output from the low-voltage DC input. If desired, a backup battery bank can keep the electric current flowing when the fuel tank is refilled. A power control

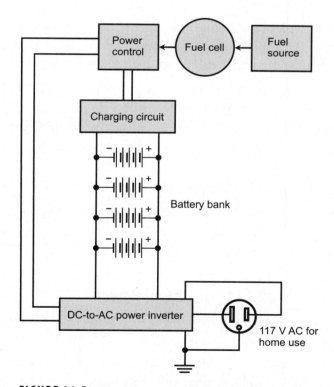

FIGURE 14-5 · Functional block diagram of a small-scale fuel cell electric power plant, suitable for use in a single residence or small business.

system switches the electrical appliances between the fuel cell and the battery bank as necessary.

TIP   *The DC output from the fuel cell or battery bank can directly provide electrical power to small appliances designed to run on low-voltage DC, such as two-way radios, some hi-fi sound systems, and all notebook computers.*

Advantages of Small-Scale Fuel-Cell Power

- Fuel cells are inherently simple, last a long time, and rarely require maintenance.
- Fuel cells offer better efficiency than conventional generators for small-scale electrification.
- Hydrogen, a favored energy source for fuel cells, produces no toxic by-products when spent.
- A properly operating hydrogen fuel cell produces negligible pollutant gases and no particulate emissions. Even if more conventional fuels, such as methane or propane, are used, less pollution will come out of a fuel cell than will come out of a combustion-engine-powered generator.
- The use of fuel cells can reduce our dependency on oil from nations that might turn hostile towards us at any time.
- The production of hydrogen for use in fuel cells could, with the proper delivery and storage infrastructure, increase the fuel supply available for heating.
- Existing pipelines can function with methane-fuel-cell power plants.
- Some workable fuels can be produced in small-scale and local facilities.
- In some locations, people who use alternative electrical energy sources, including fuel cells, can get tax breaks.

Limitations of Small-Scale Fuel-Cell Power

- In some areas, people find it difficult to get fuel cells serviced because of a lack of parts or competent technicians.
- The delivery and storage of fuel for hydrogen fuel cells presents a technological obstacle to the widespread deployment of small-scale power plants of this type. (This is not a problem with liquid alternatives.)

- The energy density of hydrogen is relatively low compared with other fuels.

- Hydrogen is potentially explosive. (This factor rarely poses a problem with liquid alternatives.)

- Hydrogen fuel cells have a relatively high operating cost, largely because of the expenses involved in separating hydrogen from naturally occurring compounds. (This factor rarely poses a problem with liquid alternatives.)

- Some fuels, such as petroleum diesel and biodiesel, tend to solidify in cold weather. This could render a fuel cell inoperative.

- Some fuels, notably methanol and gasoline, can be toxic to personnel directly exposed to them.

PROBLEM 14-3

Can we expand a system such as the one shown in Fig. 14-5 to take advantage of other sources of energy, such as sunshine or wind? If so, how might we do it? Could such a scheme allow a home to operate entirely off the electric utility grid?

SOLUTION

We can do it, but we'll probably have to spend a lot of money to make it happen. For example, a solar panel or array can charge the battery bank during sunny weather. A wind turbine can supplement the solar source, taking over on windy nights or windy, overcast days. The fuel cell can operate when neither wind nor solar energy can meet our electrical needs. A computer-governed, power-control switch can ensure that the system uses the available energy in the most efficient manner at all times. Such a hybrid system can offer complete independence from the electric utility. The key to our success in such an endeavor would lie in the diversity and redundancy of our energy sources.

Aeroelectric Power

Let's end this book by letting our imaginations run wild. Some scientists think that we might someday get power from the atmospheric electrostatic charge, which abounds on our planet. We can go all the way back to Benjamin Franklin (and some less lucky colleagues) who lofted electrical conductors into the air and demonstrated that clouds can hold a lot of "juice." Suppose that some imaginative group of engineers repeats this experiment on a grander scale, with a few refinements, and manages to get useful electricity out of it!

The Global Electric Circuit

The earth constitutes a fairly good *electrical conductor*. So does the upper part of the atmosphere known as the *ionosphere*. The lower atmosphere does not normally conduct electricity, so it constitutes an *electrical insulator*. When an insulator is sandwiched between two conductors, that insulator is known as a *dielectric*, and the result is a *capacitor* capable of storing energy as an *electric field*. On a gigantic scale it can be called a *supercapacitor*. The *earth-ionosphere super-capacitor* constantly charges up in some regions and discharges in other zones, forming a system that some scientists have called the *global electric circuit*. If humankind can ever manage to tap the global electric circuit to obtain usable electricity, we'll have invented and deployed an *aeroelectric power plant*.

How Much Power?

The maximum *electrostatic charge quantity* (number of charged particles, usually electrons or a lack of electrons) that a capacitor can hold depends on the following factors:

- The combined areas of the conducting surfaces
- The average distance separating the conducting surfaces
- The type of dielectric material between the conducting surfaces

The earth-ionosphere supercapacitor comprises one vast conducting sphere inside another with air serving as the dielectric, as shown in Fig. 14-6. These spheres both measure about 6500 km (4,000 mi) in radius. This geometry forms a capacitor with two "plates" whose spacing is relatively small (about 50 kilometers or 30 miles) compared with their surface areas (about 530,000,000 square kilometers or 200,000,000 square miles).

A high voltage between the earth's surface and the ionosphere gives rise to a massive electric field in the lower atmosphere. The charge in this supercapacitor is maintained by radiation from the sun, from cosmic rays, and from radioactivity in the earth's crust. All of this radiation interacts with the earth's magnetic field and with atoms in the upper atmosphere to keep the supercapacitor charged up.

Storm clouds, volcanoes, and dust storms temporarily improve the local conductivity of the lower atmosphere, creating attractive environments for electrical

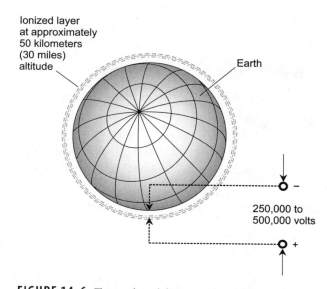

Ionized layer
at approximately
50 kilometers
(30 miles)
altitude

Earth

250,000 to
500,000 volts

FIGURE 14-6 • The earth and the upper atmosphere act as a supercapacitor that is constantly recharged by various sources of radiation.

discharge of the earth-ionosphere supercapacitor. A typical thundershower discharges about two amperes of current, averaged over time. At any given moment, roughly 750 thundershowers exist on our planet, producing between 35 and 100 lightning discharges per second altogether. A current of two amperes per thunderstorm might seem small, but this current does not flow continuously. It occurs in brief, intense surges. A single lightning discharge lasts only a few thousandths of a second. Therefore, the *peak* current in a lightning stroke can attain extreme values—in some cases many thousands of amperes. This high current explains why lightning can start fires, knock out utility systems, and kill people.

The atmospheric supercapacitor maintains a constant potential difference of several hundred thousand volts, comparable to the voltage in high-tension utility lines. It occurs as a DC voltage like a gigantic battery would produce, not an AC voltage like a utility power plant would generate. The average current that flows across the atmospheric supercapacitor as a result of thundershowers worldwide amounts to roughly 1500 amperes (two amperes per storm times 750 storms taking place at any given time).

? Still Struggling

Electrical power in watts equals the product of the voltage in volts and the current in amperes. The above-quoted figures, therefore, imply that our atmosphere constantly dissipates several hundred megawatts of power, and maybe as much as a gigawatt. That's enough to provide the electricity consumed by a medium-sized city at peak demand (but nowhere near enough to supply a nation, let alone the world).

An Aeroelectric Power Plant

How might an aeroelectric power plant work? One approach could involve lofting a set of captive high-altitude balloons tethered by conducting wires. The wires would be grounded through tanks, each containing a solution of water and dissolved electrolyte (Fig. 14-7). If such a balloon flies high enough above the surface to reach into the lowest ionized layer of the atmosphere, a constant electric current will flow in the wire and, therefore, through the electrolyte solution. This current will separate the water into hydrogen and oxygen gas, which will bubble from the electrodes. The gases would be collected in the same manner as with any other electrolysis device. The hydrogen could be used

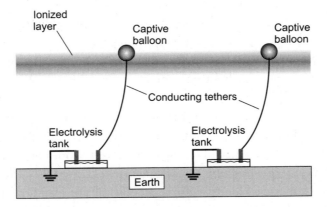

FIGURE 14-7 · A possible scheme for using atmospheric electricity to derive hydrogen fuel from water by electrolysis.

in fuel cells or hydrogen-powered cars and trucks. The oxygen could be used for industrial and medical purposes.

Advantages of Aeroelectric Power Plants

- The earth-ionosphere supercapacitor is constantly recharged by renewable energy sources, notably the sun, cosmic rays from space, and various radiation-producing elements in the earth.
- An aeroelectric power plant would produce no pollutants of any kind.
- The facilities for an aeroelectric power plant would be unobtrusive. The balloons would fly too high to be visible from the ground without binoculars or telescopes.
- The energy supply could be continuous if captive balloons were kept aloft at all times.

Limitations of Aeroelectric Power Plants

- Atmospheric electricity, like solar and wind energy, is difficult to store on a large scale. It must be used directly from the source, or else converted to some other form, such as hydrogen gas.
- If, through our activities, we discharge the earth-ionosphere supercapacitor in an unnatural way or to an unnatural extent, we might alter the balance of the global electric circuit. No one knows what environmental effects might result.
- The extreme voltages in an aeroelectric system would present a danger to technicians and other personnel working with the equipment.
- Captive balloons of the required size and altitude would be difficult to maintain and keep aloft. The balloons and conducting tethers could also present a hazard to aviation.
- The total available energy from atmospheric electricity is limited. It can, at best, serve only as a minor supplement to other energy sources.

PROBLEM 14-4

What would happen to a captive aeroelectric balloon and its associated tether if a heavy thunderstorm were to pass by, as shown in Fig. 14-8? (Let's not even think about hurricanes or tornadoes.)

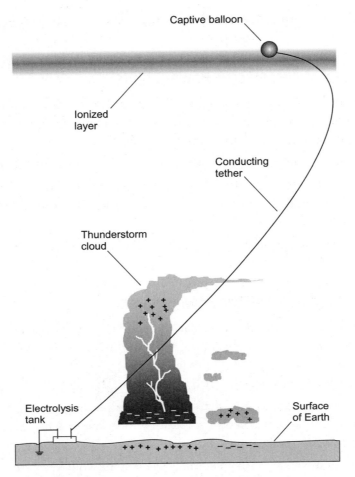

FIGURE 14-8 · What would happen if a captive aeroelectric balloon were to encounter a heavy thunderstorm?

✓ SOLUTION

The current flowing in the wire and the electrolysis tank would likely change because of localized electric charge poles in the thunderstorm. The air turbulence would stress the conducting tether. If the wire were to snap, the portion of the wire on the earth side of the break would fall to the surface near the power plant, and the portion on the balloon side would fly away with the balloon, ultimately coming back down to the surface at a distant location. This sort of event could wreak havoc with the civil infrastructure!

If humanity ever builds an aeroelectric power plant, we should expect it to be deployed on a small island in the middle of a large ocean, and there will be a backup tether or two for each balloon.

WARNING! *Don't try to deploy a small-scale model of an aeroelectric power plant at home. Even on a clear day, a wire of modest length (a couple of hundred meters) can attain a static-electric charge sufficient to generate a spark several centimeters long. In unstable weather, you could bring lightning right down to your house. If something unexpected happened to the balloon, the tether could land on a high-tension utility line, busy highway, industrial area, or some other place where it would cause a lot of trouble. Besides presenting a lethal danger to you and others, the activity would also arouse the ire of the Federal Aviation Administration (in the United States), which places tight restrictions on the use of captive balloons and large kites.*

QUIZ

Refer to the text in this chapter if necessary. A good score is eight correct. You'll find the correct answers listed in the back of the book.

1. **Aeroelectric energy shares one outstanding limitation with wind and solar energy when it comes to its viability as a reliable source of electricity. What's that limitation?**
   A. It can't work at night.
   B. It's difficult to store on a large scale.
   C. It can't work when the wind doesn't blow.
   D. It only works in storms.

2. **A geothermal power plant can be used to**
   A. directly power fuel cells.
   B. recharge the atmospheric supercapacitor.
   C. take the salt out of seawater.
   D. All of the above

3. **Some engineers have suggested methane gas (also called natural gas) as an optimal source of energy for small-scale, fuel-cell electric generating systems because**
   A. it burns more slowly than other fuels.
   B. it burns cooler than other fuels.
   C. it is less explosive than other fuels.
   D. a substantial delivery infrastructure already exists.

4. **An aeroelectric power plant that takes advantage of the entire earth's atmospheric supercapacitor (assuming that we can or will ever develop such a system) could provide enough continuous electric energy to satisfy the needs of**
   A. a small town.
   B. the entire United States.
   C. all of North America.
   D. the whole world, several times over.

5. **Which of the following locations should we reasonably expect to work best for the installation of a geothermal power plant?**
   A. The Antarctic ice sheet
   B. The slope of an active volcano
   C. The Arctic tundra
   D. The Gulf of Mexico

6. **From which of the following substances can biomass fuels be derived?**
   A. Wood
   B. Sewage

C. Microalgae

D. All of the above

7. **Which of the following substances do people often burn directly to get energy?**

A. Wood

B. Sewage

C. Microalgae

D. All of the above

8. **In a binary-cycle geothermal system, the binary liquid usually**

A. comprises oil or diesel fuel.

B. comprises biodiesel fuel.

C. resembles a refrigerant.

D. is gasoline.

9. **Biomass combustion can cause pollution by generating**

A. carbon monoxide.

B. particulate matter.

C. nitrogen oxides.

D. All of the above

10. **Which of the following factors represents a known potential problem that can occur with a poorly engineered geothermal power plant?**

A. It might cause a nearby dormant volcano to erupt.

B. Toxic gases may be released into the atmosphere.

C. It might produce too much electricity.

D. All of the above

Final Exam

Don't refer to the text when taking this test. You may draw diagrams or use a calculator if necessary. A decent score is at least 75 answers right (but why not shoot for 90?). The correct choices appear in the back of the book. I recommend that you have a friend check your results against the answer key the first time you take this test, and then tell you only your numerical score. That way, you won't "subconsciously memorize" any of the answers if you want to take the test again later.

1. If you discharge a nickel-metal-hydride (NiMH) battery all the way until it's "completely dead," you should not be surprised if that battery
   A. produces AC instead of DC the next time you charge it up.
   B. won't accept a charge again at all.
   C. works as well as ever, but its polarity has reversed.
   D. ruptures or explodes.
   E. Any of the above

2. A large wood stove, designed for use in residential homes, can provide about the same amount of heating power as
   A. a small electric space heater.
   B. several 250-watt infrared lamps.
   C. the sun shining in a small window.
   D. a good-sized gas furnace.
   E. a small solar panel.

3. Earnshaw's theorem tells us that we *cannot* obtain levitation with systems of permanent magnets
   A. made from superconducting materials.
   B. that operate from alternating current.
   C. that remain stationary relative each other.
   D. made from diamagnetic materials.
   E. at least one of which rotates.

4. On the Celsius scale, the freezing and boiling points of water differ from each other by
   A. 32 degrees.
   B. 68 degrees.
   C. 100 degrees.
   D. 180 degrees.
   E. 273 degrees.

5. Which of the following additives was once used in gasoline to prevent engine knocking, but is no longer used because it can accumulate in the environment and has proven toxic to humans?
   A. Ethanol
   B. Tetraethyl lead
   C. Petroleum diesel
   D. Methane
   E. Nitrogen tetroxide

6. Which of the following characteristics can be a disadvantage of methane as a home heating fuel?
   A. It burns inefficiently.
   B. No transport infrastructure exists.

C. It's difficult to store in residential tanks.
D. It pollutes a lot.
E. All of the above

7. **With a moderate breeze, we might expect a medium-sized wind turbine designed for residential use to provide about the same amount of electrical power as**
   A. eight electric space heaters running "full blast."
   B. a laptop computer.
   C. a large propane-fired furnace.
   D. a single generator at a typical utility power plant.
   E. several small desk lamps.

8. **Figure Exam-1 shows a cutaway view of a wood-burning stove as seen from the left-hand side of the box. What is the primary function of the component marked X?**
   A. It regulates the air intake to keep the wood burning at the desired rate.
   B. It keeps the stove from overheating by ensuring that excess hot air can vent out through the front of the stove.
   C. It ensures that the catalytic converter will work properly if the wood burns down too much.
   D. It keeps creosote from building up inside the firebox.
   E. All of the above

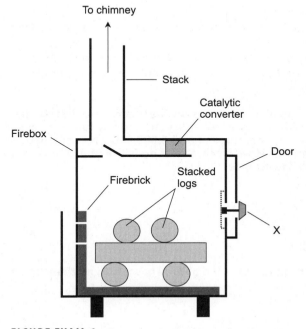

FIGURE EXAM-1 • Illustration for Final Exam Question 8.

9. If we burn a mixture of hydrogen gas and air, we get
   A. energy and nothing else.
   B. energy and water vapor, but nothing else.
   C. energy, water vapor, and a little bit of carbon monoxide gas.
   D. energy, water vapor, and a little bit of nitrous oxide gas.
   E. energy, water vapor, and a little bit of methane gas.

10. In a home designed for passive solar heating, objects warmed by the sun during the day radiate heat energy away at night in the form of
    A. ultraviolet.
    B. shortwave infrared.
    C. longwave infrared.
    D. microwaves.
    E. gamma rays.

11. In a coal-fired power plant, the coal burns in a combustion chamber after being
    A. pulverized into dust.
    B. washed with distilled water.
    C. ground up and combined with water.
    D. degassed and cut up into vitamin-pill-sized pellets.
    E. crushed into marble-sized chunks.

12. How many different major home heating appliances (furnaces or stoves) can you safely vent into a single chimney?
    A. One and only one.
    B. It depends on how many flues the chimney has. You can vent only one appliance into each flue.
    C. One or two, but never more than two at the same time.
    D. Up to three, but never more than three at the same time.
    E. As many as you want.

13. Imagine an electrical transmission line that runs for hundreds of kilometers over the countryside. It carries 400,000 volts and has 100 ohms of resistance. The end users, all taken together, demand 4,000,000 watts of electricity at a particular moment in time. How much power does this transmission line waste as heat because of its resistance?
    A. 100 watts
    B. 1000 watts
    C. 10,000 watts
    D. 100,000 watts
    E. 1,000,000 watts

14. In order to produce liquefied natural gas (LNG), methane must be
    A. distilled.
    B. cooled.
    C. boiled.
    D. refined.
    E. composted.

15. **Which of the following characteristics represents a significant advantage of LNG over compressed natural gas (CNG)?**

   A. More rapid combustion
   B. More efficient combustion
   C. Greater energy density in storage
   D. Reduced CO gas production
   E. All of the above

16. **How much energy does it take to raise the temperature of 1000 grams (1000 g) of liquid water by one degree Celsius (1°C), with none of the liquid changing state during the process?**

   A. 1 British thermal unit (1 Btu)
   B. 0.001 Btu
   C. 1 calorie (1 cal)
   D. 1 kilocalorie (1 kcal)
   E. 1000 kcal

17. **The lowest possible temperature is**

   A. 0 K.
   B. −32 K.
   C. −100 K.
   D. −273 K.
   E. None of the above

18. **The highest possible temperature is**

   A. 0 K.
   B. +32 K.
   C. +100 K.
   D. +273 K.
   E. None of the above

19. **If a forced-air heating system's *combustion* intake is located indoors, the extra air flowing into the furnace will create negative pressure inside the house,**

   A. but it won't affect the efficiency of the heating system.
   B. increasing the efficiency of the heating system.
   C. reducing the efficiency of the heating system.
   D. increasing the risk of excess oxygen buildup in the house.
   E. reducing the dust content of the air inside the house.

20. **Fill in the blank in the following statement to make it true: "In each piston of a two-stroke internal-combustion engine, ignition occurs _____, so two piston strokes occur (one down, one up) for every single ignition event."**

   A. every time the piston reaches the top of its cycle
   B. every time the piston reaches the middle of its cycle going up
   C. every time the piston reaches the middle of its cycle going down
   D. every time the piston reaches the bottom of its cycle
   E. every time the piston reaches either extreme of its cycle

21. **In theory, a diamagnetic object**
    A. attracts either pole of a permanent magnet.
    B. attracts the north pole of a permanent magnet, but repels the south pole.
    C. attracts the south pole of a permanent magnet, but repels the north pole.
    D. repels either pole of a permanent magnet.
    E. is not affected by permanent magnets in any way.

22. **How far can a typical compact electric car travel on a full charge before its battery needs recharging?**
    A. 60 miles
    B. 10 miles
    C. 600 miles
    D. 5 miles
    E. 1600 miles

23. **Fill in the blank in the following statement to make it true: "The typical RMS AC household voltage in the United States is _____ , plus or minus a few percent.**
    A. either 50 V or 100 V
    B. either 165 V or 330 V
    C. either 117 V or 234 V
    D. either 12 V or 24 V
    E. either 100 V or 200 V

24. **The pistons in a petroleum-diesel-fueled engine resemble the pistons in a gasoline-fueled engine, except for one significant difference. What's that difference?**
    A. The gasoline-fueled engine has twice as many pistons as the diesel-fueled engine has.
    B. In the gasoline-fueled engine's pistons, ignition occurs at the middle of the cycle going up, but in the diesel-fueled engine's pistons, ignition occurs in the middle of the cycle going down.
    C. The gasoline-fueled engine's pistons have spark plugs, but the diesel-fueled engine's pistons do not.
    D. The diesel-fueled engine's pistons run far hotter than the gasoline-fueled engine's pistons do.
    E. All of the above

25. **Figure Exam-2 shows a cutaway view of a sidehill home with passive solar heating. Imagine that this home is located in Santiago, Chile, at approximately 33 degrees south latitude. If this home is oriented to take maximum advantage of the sun for heating, then this drawing would have us facing**
    A. toward the west, and looking at the east side of the house.
    B. toward the north, and looking at the south side of the house.
    C. toward the east, and looking at the west side of the house.
    D. toward the south, and looking at the north side of the house.
    E. in any direction; it doesn't matter.

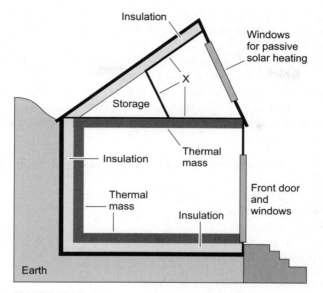

FIGURE EXAM-2 · Illustration for Final Exam Questions 25 and 26.

26. In the design of Fig. Exam-2, the best color to paint the wall, floor, and ceiling marked X would be

    A. white.
    B. gray.
    C. black.
    D. red.
    E. blue.

27. For an alternative liquid motor-vehicle fuel, the ratio of the energy available in a certain volume of conventional gasoline to the energy available in the same volume of the alternative fuel (using identical energy units for both fuels) is called the

    A. gasoline-gallon equivalent.
    B. joule mileage ratio.
    C. mileage per volume coefficient.
    D. energy density factor.
    E. power production metric.

28. Modern oil-fired power plants produce less pollution than they did decades ago, primarily because the newer plants take advantage of

    A. fracking.
    B. emission-control apparatus.
    C. de-gassed oil.
    D. lower combustion temperatures.
    E. All of the above

29. Suppose that you decide to install a medium-sized residential wind turbine for your home. Which of the following considerations is a significant disadvantage of this type of system?

    A. You won't be able to use it in any system that includes storage batteries.
    B. Your neighbors might object because it would ruin their view.
    C. It will generate significant carbon dioxide, a known greenhouse gas.
    D. It will need at least a 25-mile-an-hour wind to work properly.
    E. You'll never be able to use it in any system that interconnects with the electric utility.

30. In which of the following devices or machines should you *never* expect to find a lead-acid battery?

    A. An uninterruptible power supply
    B. A gasoline-powered car
    C. A notebook computer
    D. A proton-exchange-membrane (PEM) fuel cell
    E. A wood-pellet or corn-burning stove

31. Hydrogen combustion does not produce

    A. carbon compounds.
    B. methane gas.
    C. carbon monoxide gas.
    D. particulate pollution.
    E. Any of the above

32. How do some small-scale generators for residential use circumvent the need for constant motor speed?

    A. They use special transformers with adjustable windings to keep the AC frequency at a constant 60 Hz.
    B. They convert the generated AC to regulated DC, and then use power inverters to generate clean 60-Hz AC from that DC.
    C. They prevent the load from varying significantly, so the motor does not experience any change in shaft resistance.
    D. They use special electromechanical devices called back-pressure sensors to keep the motor shaft resistance constant.
    E. They don't because they can't.

33. A diversion type hydroelectric power plant can comprise a

    A. reservoir and a dam on a river.
    B. chamber to trap turbulent ocean water.
    C. device to contain water from littoral currents.
    D. system to capture wave energy.
    E. water turbine in a pipeline next to a fast-moving stream.

34. The heating elements usually reside near the ceiling in a

    A. radiant electric zone heating system.
    B. hot-water heating system.

C. steam heating system.

D. forced-air heating system.

E. photovoltaic heating system.

35. **A grid-intertie, wind-power system for a residence or small business might include any or all of the following devices** *except one*. **Which one?**

A. A system that allows for selling excess power to the electric utility.

B. A penstock to direct the wind through the turbine.

C. A device that allows for buying power from the utility.

D. A power inverter to convert low-voltage DC to utility AC.

E. A storage battery to provide power when the wind does not blow.

36. **When a nuclear fission reactor operates in the subcritical state,**

A. the reaction dies down before much fuel is spent.

B. then on the average, one emitted neutron strikes another uranium-235 (U-235) nucleus and splits it.

C. neutrons never strike any U-235 nuclei.

D. hydrogen atoms combine to form helium atoms at a regular, controlled rate.

E. then on the average, more than one emitted neutron strikes another U-235 nucleus and splits it.

37. **Figure Exam-3 is a block diagram of a small water-driven energy system adapted for electric baseboard home heating. The box marked X represents an essential component of such a system. That component is a**

A. storage battery.

B. power inverter.

C. voltage regulator.

D. switch.

E. fuel cell.

38. **In order for a nuclear fission reactor to work properly, it must be maintained in a condition such that**

A. the reaction dies down before much fuel is spent.

B. on the average, one emitted neutron strikes another U-235 nucleus and splits it.

C. neutrons never strike any U-235 nuclei.

D. hydrogen atoms combine to form helium atoms at a regular, controlled rate.

E. on the average, more than one emitted neutron strikes another U-235 nucleus and splits it.

39. **In order to generate electricity, a nuclear fusion reactor would theoretically make use of deuterium (which can be obtained from water), lithium (abundant in the earth's crust), and**

A. beryllium (which can be derived as a by-product of coal refining).

B. uranium (which can be mined conventionally).

C. high-speed neutrons (which can be generated by a fission reaction).

D. tritium (which can be created in the reactor itself).

E. oxygen (which can be derived from the atmosphere).

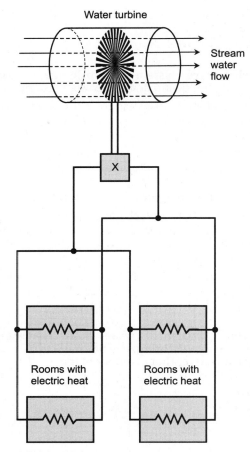

Water turbine

Stream
water
flow

X

Rooms with
electric heat

Rooms with
electric heat

FIGURE EXAM-3 • Illustration for Final Exam
Question 37.

40. **If a forced-air heating system is used in a house that's completely airtight, fresh outdoor air won't be able to get into the house,**
    A. and this effect will reduce the danger of an explosion.
    B. eventually causing all the oxygen in the air to deplete.
    C. reducing the efficiency of the heating system.
    D. increasing the risk of carbon-monoxide buildup in the house.
    E. so dust will eventually build up to intolerable levels.

41. **The current that discharges in the atmospheric supercapacitor, as averaged over time, equals roughly**
    A. 15 amperes.
    B. 150 amperes.

C. 1500 amperes.

D. 15,000 amperes.

E. 150,000 amperes.

42. **You might find a halogen bulb employed in all of the following applications except one. Which one?**

A. A vehicle headlight

B. An art gallery

C. A "neon sign"

D. A "movie" projector

E. A desk lamp

43. **A small-scale, hydroelectric power plant, intended for a single residence located by a fast-moving stream, would most likely take advantage of**

A. wave energy.

B. diversion technology.

C. impoundment technology.

D. tidal currents.

E. littoral currents.

44. **Which of the following statements is false?**

A. CFLs lose a little bit of their brilliance as they age.

B. If you break a CFL, you must take special precautions to clean up the mess.

C. If the temperature falls far below freezing, CFLs might not start up.

D. CFLs produce more light per unit surface area than any other lamp type.

E. CFLs can sometimes interfere with wireless devices.

45. **You can eliminate your need for any major home cooling or heating system, even if the outdoor climate is extreme, by taking advantage of**

A. grid-intertie wind turbines.

B. stand-alone photovoltaic power.

C. the subterranean living option.

D. evaporative power technology.

E. tidal power.

46. **Fill in the blank in the following statement to make it true: "A properly operating propane-fueled vehicle engine produces _____ than a properly operating gasoline- or diesel-powered engine of the same size."**

A. more toxic by-products

B. greater energy density

C. more noise and knock

D. less air pollution

E. greater dependency on foreign oil

47. One of the following statements concerning light-emitting-diode (LED) computer or television displays is false. Which one?

A. They cost less to operate, in terms of electric bills, than cathode-ray-tubes (CRTs) of the same size do.
B. They cost a lot of money at the store, relative to older display types.
C. They produce brilliant, color-faithful images.
D. They respond rather slowly, so they won't work very well for fast-changing images or videos.
E. They are semiconductor-based devices.

48. We can obtain perfect diamagnetism if we can get a sample of material to

A. concentrate magnetic lines of flux to an infinite extent.
B. act as a superconductor.
C. accommodate an alternating magnetic field.
D. behave as a powerful enough electromagnet.
E. do all of the foregoing things at the same time.

49. One of the main problems with basic passive solar heating, where you simply let the sun shine in through large windows during daylight hours, is color fading of wallpaper, carpeting, and furniture. The type of energy most responsible for this fading is ultraviolet (UV). You can minimize the intrusion of UV by

A. keeping the furnace or air-conditioner fan running all the time.
B. installing double-pane or triple-pane window glass.
C. painting the ceiling white, or putting white acoustical tile on it.
D. switching off the furnace or air-conditioner fan during the daytime.
E. Any of the above

50. The earth-ionosphere supercapacitor is constantly recharged by renewable energy sources including

A. electricity generated by humans.
B. oceanic wave energy.
C. cosmic rays from space.
D. oceanic tide energy.
E. All of the above

51. With respect to wind turbines, a 50-mile-an-hour wind has

A. the square root of 2 times the theoretical power of a 25-mile-an-hour wind.
B. twice the theoretical power of a 25-mile-an-hour wind.
C. four times the theoretical power of a 25-mile-an-hour wind.
D. eight times the theoretical power of a 25-mile-an-hour wind.
E. 16 times the theoretical power of a 25-mile-an-hour wind.

52. Figure Exam-4 is a diagram of an electrochemical cell designed to produce low-voltage DC electricity. What substance is represented by the X?

A. Distilled water
B. Sulfuric acid

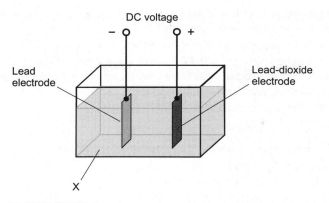

FIGURE EXAM-4 · Illustration for Final Exam Question 52.

C. Hydrogen peroxide
D. Salt water
E. Potassium hydrochloride

53. **In a flash-steam geothermal power plant, the turbines are driven *directly* by**
    A. flowing liquid water.
    B. air under pressure.
    C. water vapor under pressure.
    D. pressurized coolant.
    E. nothing, because this type of system has no turbines.

54. **In a thermal-mass cooling system, the exterior walls should be painted white on the outside, or covered with white siding, in order to**
    A. keep the sun's UV rays from fading them (a white surface can't get any lighter than it already is).
    B. maximize the speed with which they radiate heat away at night.
    C. prevent excessive thermal conduction between the home's interior and the outside air.
    D. minimize the heat that they absorb from sunshine during the day.
    E. facilitate vertical convection along the outside of the structure.

55. **Which of the following fuel types can crystallize at frigid temperatures such as might be encountered in an Arctic climate?**
    A. Methane
    B. Hydrogen
    C. Gasoline
    D. Diesel
    E. Ethanol

56. **Which of the following characteristics** *does not* **represent a characteristic of ethanol for use as an additive to gasoline?**

    A. Ethanol can serve as an extender and octane enhancer in conventional gasoline.
    B. Ethanol can help prevent gas-line freeze in extremely cold weather.
    C. The plant matter used to produce ethanol constitutes a renewable resource.
    D. Ethanol is more flammable than gasoline.
    E. Ethanol can reduce emissions of deadly CO gas.

57. **Which of the following statements applies to coal stoves?**

    A. Coal is not as clean-burning as natural gas.
    B. A coal stove needs to be cleaned often.
    C. Coal can be hard to get in some locations.
    D. Some local governments forbid coal-burning stoves.
    E. All of the above

58. **In a maglev train, acceleration and deceleration are provided by**

    A. linear motors.
    B. diamagnetic tracks.
    C. conventional gasoline engines.
    D. wheels between the cars and the track.
    E. superconducting permanent magnets.

59. **Which of the following statements is** *false* **with regard to an interactive small-scale home photovoltaic (PV) system with batteries?**

    A. If there's a prolonged spell without enough light for the PV cells to function, the electric utility can keep the batteries charged.
    B. This type of system can deliver more peak power, in general, than a similar system that does not have batteries.
    C. A switch, along with a battery-charge detection circuit, connects the batteries to the utility through a charger if insufficient power comes from the PV array.
    D. When conditions become favorable, a switch disconnects the batteries from the utility charger and connects them to the PV array.
    E. Power is never sold back to the electric utility, even when the PV panel or array generates more power than the home needs.

60. **Although several thousand DC volts can theoretically be obtained by connecting a great many PV cells in series, this approach doesn't work well in practice. Why?**

    A. Such a system will provide too much current for most electrical grids to safely handle.
    B. The individual PV cells will tend to burn out frequently because of the extreme voltage to which they get subjected.

C. The internal resistances of the cells will add up, causing the voltage to drop significantly under load.

D. One or more of the PV cells can, and quite likely will, reverse polarity if the system is kept under load for too long. That condition will cause the whole system to fail.

E. The premise of this question is wrong! Many PV cells can be connected together in series to get several thousand volts, and such a system will work just fine.

61. **Air-source heat pumps can heat the indoor air quite well, as long as the outdoor temperature remains above about**

    A. 0 K.
    B. 0°F.
    C. +4°C.
    D. +72°F.
    E. 212 K.

62. **Which of the following statements represents a major advantage of conventional jet fuel for aircraft propulsion?**

    A. Its combustion doesn't produce any sulfur byproducts.
    B. Its combustion doesn't produce any carbon dioxide.
    C. No one has found anything better yet.
    D. The fuel doesn't have to come from oil or coal.
    E. All of the above

63. **Which of the following characteristics represents a significant limitation or disadvantage of solar water heating?**

    A. The electrical pump uses a lot of electricity, and it tends to fail often because of the hot water it must handle.
    B. It's a closed system, so it can't ventilate efficiently if the weather gets too cold.
    C. It won't work very well in a location where the weather is cloudy most of the time.
    D. It generates pollutants in the form of dissolved metals from the pipe walls; these metals must eventually be discarded somehow, so they inevitably end up in the environment.
    E. All of the above

64. **Figure Exam-5 is a simplified block diagram of a nuclear power and propulsion system suitable for use in**

    A. propeller-driven aircraft.
    B. predator drones.
    C. submarines.
    D. experimental high-speed overland vehicles.
    E. Any of the above

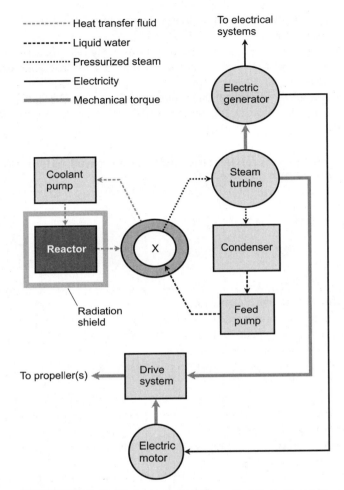

FIGURE EXAM-5 · Illustration for Final Exam Questions 64 and 65.

65. **In Fig. Exam-5, the ellipse at the center, marked with an X, represents a**
    A. heat pump.
    B. boiler.
    C. condenser.
    D. combustion chamber.
    E. tokamak.

66. **Fill in the blank to make the following sentence true: "It will take _____ energy to produce a cubic meter of hydrogen by water electrolysis (at standard atmospheric pressure) as we will get when we burn that same cubic meter of hydrogen."**
    A. half as much
    B. two-thirds as much

C. half again as much
D. the same amount of
E. three times as much

67. **Which of the following statements** *does not* **describe an advantage of methane for use in generating electricity?**

 A. Methane is readily available in most cities and towns, near the end users of the electricity.
 B. Methane-fired, combined-cycle power plants are efficient.
 C. Methane-fired power plants might be modified to burn hydrogen gas if that fuel source becomes available in quantity at a reasonable price.
 D. Exploration for, and recovery of, methane has no known adverse impact on the environment.
 E. An uninterrupted supply of methane can be provided by underground pipelines, reducing the need for energy-consuming trains and trucks to transport the fuel from the refinery to the electric-generating plant.

68. **Composting can be used on a small scale to produce**

 A. methane.
 B. propane.
 C. biodiesel.
 D. ethanol.
 E. gasohol.

69. **Which of the following statements** *does not* **represent a significant limitation of, or problem with, electric-vehicle (EV) technology?**

 A. An EV might not provide enough power to operate in heavy snow, or to carry heavy loads on bad roads.
 B. An EV actually produces more pollution overall, because of the nature of the manufacturing process and because of the sources of its charging power, than a gasoline-powered vehicle of comparable size.
 C. The battery in an EV can lose some of its ability to hold a charge if the outdoor temperature falls below freezing.
 D. It can be hard to get an EV serviced because of a lack of available parts or competent technicians.
 E. Drivers must use extra caution to stay safe, because EVs are smaller than most other vehicles on the road.

70. **Figure Exam-6 is a simplified diagram of a**

 A. stand-alone photovoltaic (PV) system.
 B. grid-intertie PV system with batteries.
 C. utility-dependent PV system.
 D. redundant PV system.
 E. hybrid PV system.

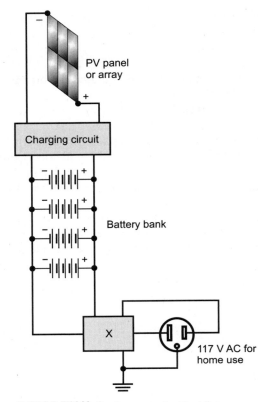

FIGURE EXAM-6 · Illustration for Final Exam Questions 70 and 71.

71. **In the system of Fig. Exam-6, the component marked X is technically called**
    A. a transformer.
    B. a filter.
    C. a power inverter.
    D. an uninterruptible power supply.
    E. a voltage regulator.

72. **It takes a certain amount of energy to change a sample of liquid to its gaseous state, assuming the matter can exist in either of these two states. Scientists call it the**
    A. condensation coefficient.
    B. coefficient of performance.
    C. vaporization point.
    D. heat of vaporization.
    E. energy conversion coefficient.

73. **With a direct PV power system designed for home cooling, you should be able to power all of the following devices except one. Which one?**
    A. An electric fan
    B. A central air-conditioning unit
    C. A humidifier
    D. An evaporative cooler
    E. None of the above; a direct PV system can power them all.

74. **A modern coal-fired power plant can burn**
    A. anthracite only.
    B. lignite only.
    C. bituminous only.
    D. all grades of coal.
    E. oil derived from coal, but not the coal itself.

75. **Which of the following characteristics *does not* describe an advantage of large wind-power-based electric generation systems?**
    A. Wind power plants do not directly pollute the environment.
    B. Large wind turbines can be placed over the ocean, as well as on land.
    C. Wind is a renewable energy source, and the supply is unlimited.
    D. A single massive wind turbine can continuously supply all the power that a small community needs.
    E. Once installed, a large wind turbine is not particularly expensive to maintain.

76. **Halogen lamps are known for the fact that they**
    A. can produce brilliant light.
    B. are more efficient than any other lamp type.
    C. produce very little heat.
    D. cost more to buy than any other lamp type of the same wattage.
    E. contain mercury, a known toxin.

77. **All of the following materials have good thermal mass characteristics except one. Which one?**
    A. Brick
    B. Adobe
    C. Concrete
    D. Wood
    E. Granite

78. **Figure Exam-7 is a simplified functional diagram of a**
    A. binary-cycle geothermal power plant.
    B. magma-driven geothermal power plant.
    C. liquid-water-driven geothermal power plant.
    D. triple-cycle geothermal power plant.
    E. fuel-cell geothermal power plant.

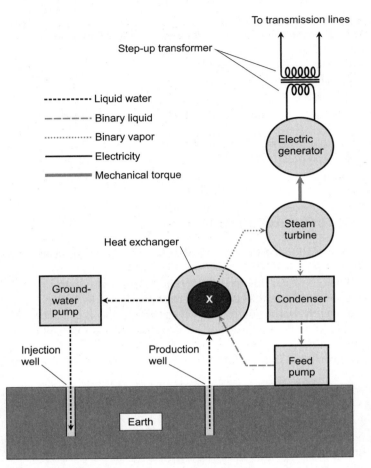

FIGURE EXAM-7 · Illustration for Final Exam Questions 78 and 79.

79. **In the system of Fig. Exam-7, the component marked X is a**

    A. magma distiller.
    B. fuel cell.
    C. flash tank.
    D. boiler.
    E. coolant tank.

80. **The course or motion of a space ship using a solar sail might be affected by**

    A. interstellar nebulae.
    B. irregularities in the fission process.
    C. intense magnetic fields near certain planets.

D. ultraviolet radiation from the sun.

E. Any of the above

81. **In any grid-intertie, residential electrical system that shares wiring between a wind turbine and the utility grid, precautions must be taken to prevent**

A. the wind turbine from generating more power than the home needs.

B. the turbine's AC from getting out of phase with the utility's AC.

C. the wind turbine from becoming disconnected from the system.

D. the batteries from overcharging or totally depleting.

E. Any or all of the above

82. **An impoundment type hydroelectric power plant can incorporate a**

A. reservoir and a dam on a river.

B. chamber to trap turbulent ocean water.

C. device to contain water from littoral currents.

D. system to capture wave energy.

E. water turbine in a pipeline next to a fast-moving stream.

83. **We can express accumulated power over a period of time in**

A. British thermal units.

B. joules.

C. calories.

D. kilowatt-hours.

E. Any of the above

84. **Which of the following substances *never* serves as part of the propellant (either the combustible fuel or its oxidizer) in a liquid-fueled rocket engine intended for transportation in outer space?**

A. kerosene

B. hydrazine

C. liquefied carbon dioxide

D. liquefied oxygen

E. nitrogen tetroxide

85. **In a dry-steam geothermal power plant,**

A. water is not needed at all; hot air from deep under the ground comes up under pressure and drives the turbines.

B. the water vaporizes on its way up through the production well, eliminating the need for a flash tank.

C. heat from uranium fission reactions adds to the natural geothermal heat, producing superheated steam.

D. steam comes up directly from far beneath the surface, in much the same way as it would in a naturally occurring geyser.

E. the water returns to the liquid state and gets forced by the groundwater pump back down into the earth along with the diverted water from the flash tank.

86. **Compact fluorescent lamps (CFLs) are known for the fact that they**
    A. produce relatively little light in general.
    B. are less efficient than any other lamp type.
    C. present a cleanup inconvenience if you break them.
    D. produce almost no heat.
    E. cost less to buy than any other lamp type of the same wattage.

87. **Which of the following characteristics constitutes an advantage of a hot-water home heating system?**
    A. It's a closed system, so it's maximally efficient.
    B. Negative pressure doesn't occur inside the house.
    C. It won't drive outdoor dust into the house.
    D. It lends itself to radiant heat subflooring.
    E. All of the above

88. **Fill in the blank to make the following statement correct: "In an electric vehicle (EV) that takes advantage of regenerative braking, a _____ can harness some of the energy that would otherwise go to waste heating the brake drums, and use that energy to recharge the battery."**
    A. photovoltaic panel and power inverter
    B. proton-exchange-membrane (PEM) fuel cell
    C. generator and alternating-to-direct-current (AC/DC) converter
    D. small gasoline-powered engine
    E. water-electrolysis system

89. **Which of the following characteristics constitutes an advantage of thermal-mass heating?**
    A. Thermal mass reduces fluctuations in temperature between day and night, or between sunny days and cloudy days.
    B. If a prolonged cloudy spell occurs, the heat acquired during sunny periods will last longer in a building with more thermal mass than in a building with less.
    C. The inclusion of thermal mass in new construction can result in a building more likely to withstand severe weather, particularly high winds accompanied by flying debris.
    D. A building constructed with substantial concrete, stone, or brick offers better resistance to fire than a conventional frame building does.
    E. All of the above

90. **Multiple reservoirs whose water levels have different elevations are used in a**
    A. diversion type hydroelectric power plant.
    B. penstock-release type hydroelectric power plant.
    C. pumped-storage type hydroelectric power plant.
    D. littoral-current type hydroelectric power plant.
    E. tidal-current type hydroelectric power plant.

91. **In a small-scale fuel-cell power plant designed for residential use, we might, at least in theory, use the output of the fuel cell to *directly* provide the energy to operate**

  A. a notebook computer.
  B. an electric space heater.
  C. an electric oven.
  D. an electric air conditioner.
  E. Any of the above

92. **Methane is generally safer than propane as a heating fuel because**

  A. methane is more energy-dense than propane.
  B. methane burns cooler than propane.
  C. methane is less explosive than propane.
  D. propane can freeze solid but methane can't.
  E. The premise of this question is wrong. Propane is actually safer, in general, than methane.

93. **Dilute acid esterification is a process commonly used in the production of**

  A. gasoline.
  B. biodiesel.
  C. propane.
  D. fuel cells.
  E. refrigerants.

94. **When talking about the operation of a heat pump, engineers sometimes refer to the ratio of the thermal energy actually transferred, to the input energy that the system needs to do the job. The technical term for this specification is**

  A. heat entropy.
  B. thermal efficiency.
  C. coefficient of performance.
  D. heat of vaporization.
  E. coefficient of mobilization.

95. **So-called black powder, used as a solid fuel in some military rockets, can be made with**

  A. carbon and other solid waste products from diesel-fuel combustion.
  B. charcoal, potassium nitrate, and sulfur.
  C. carbon powder mixed with water and then dried into a hard mass.
  D. crystallized, charred, compressed biodiesel fuel.
  E. ordinary table sugar, charred and then compressed.

96. **Figure Exam-8 is a diagram of a**

  A. conventional fluorescent tube.
  B. neon lamp.
  C. halogen lamp.
  D. cathode-ray tube.
  E. light-emitting diode.

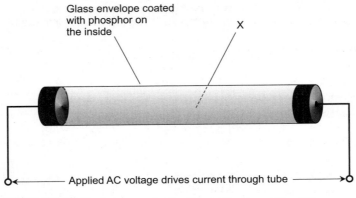

FIGURE EXAM-8 · Illustration for Final Exam Questions 96 and 97.

97. **The interior of the device in Fig. Exam-8, marked X, contains**
    A. neon gas.
    B. halogen gas.
    C. a vacuum.
    D. mercury vapor and an inert gas.
    E. a semiconducting medium.

98. **An infrared lamp transfers heat energy mainly by**
    A. radiation.
    B. convection.
    C. conduction.
    D. evaporation.
    E. condensation.

99. **We should expect an evaporative indoor weather modification system to work well in**
    A. a hot, wet place such as the Amazon river basin in Brazil.
    B. a cold, snowy place such as Aspen, Colorado in the winter.
    C. a warm, dry place such as Reno, Nevada in the summer.
    D. a cold, dry place such as Laramie, Wyoming in the winter.
    E. Any of the above

100. **If a propane home heating system springs a significant fuel leak,**
    A. you'll always hear a loud hissing noise.
    B. you'll sense a decrease in indoor humidity.
    C. you'll probably detect a rotten-egg-like odor.
    D. your house will probably get too hot.
    E. Any or all of the above

Answers to Quizzes and Final Exam

Chapter 1	Chapter 3	Chapter 5	Chapter 7
1. C	1. B	1. B	1. B
2. A	2. D	2. D	2. B
3. D	3. A	3. D	3. C
4. B	4. C	4. B	4. D
5. B	5. B	5. B	5. C
6. A	6. B	6. D	6. B
7. B	7. A	7. C	7. A
8. C	8. D	8. A	8. D
9. D	9. C	9. C	9. B
10. A	10. B	10. A	10. A

Chapter 2	Chapter 4	Chapter 6	Chapter 8
1. D	1. D	1. A	1. B
2. C	2. D	2. D	2. D
3. A	3. A	3. C	3. B
4. C	4. B	4. C	4. B
5. D	5. A	5. C	5. D
6. C	6. C	6. B	6. C
7. B	7. C	7. A	7. C
8. C	8. D	8. D	8. A
9. A	9. B	9. B	9. A
10. D	10. B	10. D	10. D

Chapter 9
1. B
2. C
3. A
4. C
5. C
6. D
7. A
8. B
9. A
10. C

Chapter 10
1. C
2. D
3. B
4. D
5. A
6. B
7. B
8. A
9. D
10. C

Chapter 11
1. B
2. B
3. A
4. C
5. C
6. A
7. D
8. B
9. B
10. D

Chapter 12
1. D
2. B
3. B
4. D
5. D
6. C
7. A
8. C
9. A
10. D

Chapter 13
1. B
2. A
3. A
4. C
5. B
6. C
7. C
8. D
9. B
10. D

Chapter 14
1. B
2. C
3. D
4. A
5. B
6. D
7. A
8. C
9. D
10. B

Final Exam
1. B
2. D
3. C
4. C
5. B
6. C
7. A
8. A
9. D
10. C
11. A
12. B
13. C
14. B
15. C
16. D
17. A
18. E
19. C
20. A
21. D
22. A
23. C
24. C
25. A
26. C
27. A
28. B
29. B
30. D
31. E
32. B
33. E
34. A
35. B
36. A
37. C
38. B
39. D
40. D
41. C
42. C
43. B
44. D
45. C
46. D
47. D
48. B
49. B
50. C
51. D
52. B
53. C
54. D
55. D
56. D
57. E
58. A
59. B
60. C
61. C
62. C
63. C
64. C
65. B
66. C
67. D
68. A
69. B

70. A	78. A	86. C	94. C
71. C	79. D	87. E	95. B
72. D	80. C	88. C	96. A
73. B	81. B	89. E	97. D
74. D	82. A	90. C	98. A
75. D	83. E	91. A	99. C
76. A	84. C	92. E	100. C
77. D	85. B	93. B	

Suggested Additional Reading

Chiras, Daniel D., *The Homeowner's Guide to Renewable Energy: Achieving Energy Independence through Solar, Wind, Biomass, and Hydropower*, 2nd ed. Gabriola Island, BC, Canada: New Society Publishers, 2011.

Chiras, Daniel D., *The Solar House: Passive Heating and Cooling*. White River Jct., VT: Chelsea Green, 2002.

DeGunther, Rik, *Alternative Energy for Dummies*. Hoboken, NJ: Wiley Publishing, Inc., 2009.

Ewing, Rex A. and LaVonne, *Crafting Log Homes Solar Style: An Inspiring Guide to Self-Sufficiency*. Masonville, CO: PixyJack Press, 2008.

Ewing, Rex A. and Pratt, Doug, *Got Sun? Go Solar*, 2nd ed. Masonville, CO: PixyJack Press, 2009.

Gibilisco, Stan, *Physics Demystified*, 2nd ed. New York, NY: McGraw-Hill, 2011.

Gipe, Paul, *Wind Energy Basics*, 2nd ed. White River Jct., VT: Chelsea Green, 2009.

Hodge, B. K., *Alternative Energy Systems*. Hoboken, NJ: Wiley Publishing, Inc., 2009.

Kachadorian, James, *The Passive Solar House*. White River Jct., VT: Chelsea Green, 2006.

MacKay, David, *Sustainable Energy Without the Hot Air*. Cambridge, UK: UIT Cambridge Ltd., 2009.

Index